John S. Rinehart

Geysers and Geothermal Energy

Springer-Verlag
New York Heidelberg Berlin

John S. Rinehart
P.O. Box 392
Santa Fe, NM 87501
USA

Library of Congress Cataloging in Publication Data

Rinehart, John Sargent, 1915−
  Geysers and geothermal energy.

  Bibliography: p.
  Includes index.
  1. Geysers. 2. Geothermal resources.
I. Title
QE528.R56   551.2′3   80-14301

With 97 figures.

Printed in the United States of America.

9 8 7 6 5 4 3 1

ISBN 0-387-90489-1 Springer-Verlag New York
ISBN 3-540-90489-1 Springer-Verlag Berlin Heidelberg

Dedicated to my wife Marion

Preface

Geysers. What makes them work? Many who have seen a geyser in action know only that it spouts hot water into the air. Many others have never seen one. Chapter 1, Geysers of the World, delineates their distinguishing features, locates the geyser regions of the world, and places investigations by world travelers and scientists in historic perspective. One of the quickest ways to become acquainted with a geyser is to observe it. The descriptions of several well known geysers, some based on past observations by others, but frequently by me, do not necessarily portray current behavior. They do, however, represent general features.

Geysers exist as a result of a delicate and unique interplay among the heat, the water, and the rocks of the earth. In essence, heat and water must be available, transported, distributed, stored, and finally released. Chapter 2, The Geologic, Thermal, and Hydrologic State of the Earth, especially that close to its surface, sets the stage for Chapter 3, Fundamentals of Geyser Operation. The geyser is treated here as a simple system consisting of three major interacting elements: a source of water, a source of heat, and a reservoir for storing water. The discussion centers around the actions occurring within idealized columnar and pool geysers, and more complex systems. Some of the more workable geyser theories are evaluated.

Gases and minerals carried by thermal waters play very important roles in many geologic processes. In geysers, gases strongly affect the eruption processes. Minerals do not, but the chemical composition of the waters provides a key to understanding where the waters have come from and been. Whereas Chapter 3 ignores the presence of gases as an agent in geyser action, Chapter 4, The Role of Gases in Geysers, discusses their importance in detail. Especially interesting are the cold water geysers whose eruptions are powered mainly by occluded and dissolved gases other than water vapor. Chapter 5, Chemistry of Geothermal Waters, discusses the chemical characteristics of geyser waters, their

origins, and their passages through rock masses. Minerals are essential in the formation and maintenance of geyser systems by solution of the subterranean rocks through whose channels the waters can circulate, and by deposition which forms the underground reservoirs, auxiliary tubes, and surface appurtenances essential to the operation of a geyser.

Chapter 6 is concerned with Geyser Area Complexes, the differences and interrelationships between mud pots, fumaroles, spouters, nonerupting hot springs, and geysers.

Hot spring and geyser activity have strong influences on local environments. The hot waters provide habitats for out-of-the-ordinary plant and animal life. A geyser's steam explosions can shake the ground. Because of their fragile character, unpredictable behavior, and masses of boiling water, hot spring areas are often hazardous to human and animal life. Chapter 7 is concerned with these Environmental Aspects of Geysers.

Geyser activity is constantly changing, at times secularly, at times episodically, and often cyclically. Many factors cause these changes: climate, air pressure, earthquakes and their associated earth stresses, and earth tides. An eruption pattern also has vagaries that seem characteristic of a geyser system itself. Chapter 8 details all of these Temporal Changes in Geyser Activity and Their Causes.

It is natural for man to meddle with geysers, trying to change or understand their behavior. Indeed he has revived inactive geysers, stimulated sluggish ones, and made new ones. At time he has harnessed them for useful work. Chapter 9 is about Man's Influence on Geyser Activity.

Chapter 10 is concerned with Practical Uses of Geothermal Fluids. Geothermal fluids, hot water and steam, contain vast amounts of energy. The fluids are now used in appreciable quantities for electric power generation, space heating, industrial processing, and agriculture. All of these uses and the types of geothermal areas that can be effectively utilized are discussed.

This book brings together most aspects of geyser activity. It differs from past discussions, which though extensive and excellent have either been guide books or limited expositions of a single geyser area or phenomenon. Here information from all of the geyser areas of the world is used to establish the causes, nature, and effects of geyser activity.

In preparing this book, available geyser literature has been drawn on heavily, often simply paraphrased. Only figures and tables are referenced specifically. All material consulted is listed in the bibliography with each chapter's references identified. All of the major works contain additional and sometimes extensive bibliographies for further study.

I have visited most of the areas discussed with the exception of Kamchatka and all of the photographs are my own. During these visits and at other times, I have benefited greatly from discussions and personal help from D. W. White, R. O. Fournier, A. H. Truesdell, George Marler, R. W. Hutchinson, Joe Barth, and E. F. Lloyd. Vincent Schaeffer kindled my scientific interest in geysers. My many students in the Mechanical Engineering Department at the University of Colorado have been invaluable in analysing the data. I am also grateful to the

naturalist-rangers and officials of Yellowstone National Park who have been so very cooperative during my stays there.

My wife has been a continual source of help and inspiration. She has been my companion and technician on many arduous expeditions and has participated in all aspects of my writing including typing and editing of the several drafts of this manuscript.

Santa Fe, New Mexico   JOHN S. RINEHART

Contents

Chapter 10
Practical Uses of Geothermal Fluids

List of Symbols

A	Area
A_b	Area of bubble
c	Specific heat of water
C_{T_a}	Mineral concentration at temperature in aquifer
d	Depth within geyser tube (top down to point under consideration)
Δd	Change in distance boiling surface has moved
d_e	Depth at which boiling reaches equilibrium
d_g	Depth at which boiling begins (gas-filled water)
Δd_g	Distance ebullition of gas works itself down to equilibrium
d_1	Depth at which boiling begins (no superincumbent gas)
Δd_1	Distance boiling surface has moved downward from initial location to equilibrium depth
D	Length of geyser tube
E	Enthalpy
E_1	Enthalpy of state one
E_2	Enthalpy of state two
ΔE	Change in enthalpy
F	Force of attraction
g	Acceleration of gravity
Δg_r	Change in vertical component in gravity
Δg_ψ	Change in horizontal component in gravity
G	Universal gravitational constant

h	Height
K	(1) Coefficient of permeability; (2) Henry's constant (amount of gas in solution in water at given temperature and pressure); and (3) thermal conductivity
KE	Kinetic energy
l	Length of borehole
m	Mass of water in vent of Velikan just before eruption
m_b	Mass of bubble wall per gram of steam
m_s	Mass of steam discharged by Velikan during play
m_w	Mass of water discharged by Velikan during play
m_T	Total mass discharged by Velikan
m_1	Mass of water and steam emptied from Velikan
M_E	Mass of earth
M_M	Mass of moon
M_S	Mass of sun
M_1, M_2	Mass of bodies 1 and 2, respectively
N	Number of bubbles
p	(1) Vapor pressures; (2) Pressure on water at depth d
Δp	Pressure difference
p_{atm}	Standard atmospheric pressure
p_o	Ambient atmospheric pressure
p_{G_1}	Vapor pressure at depth d_1
p_2	Pressure in reservoir
P	Hydrostatic pressure
P_1	Hydrostatic pressure at depth d_1
q	(1) Rate of heat transfer; (2) Rate of inflow of fluid
q_c	Rate of inflow of cold water
q_h	Rate of inflow of hot water
q_w	Rate of inflow of water into geyser tube per unit cross section of tube area
q_1	Total rate of inflow, $q_c + q_h$
Q_{w+s}	Rate of inflow of water and steam
r	Radius
R	Separation of body masses
R_M	Distance between earth and moon
R_S	Distance between earth and sun

t	Temperature, °C
t_w	Temperature at which water enters geyser tube
t_1	Temperature at which water enters geyser tube exceeds boiling point at surface
T	Absolute (Kelvin) temperature
ΔT	Temperature difference
T_a	Temperature in aquifer
T_c	Temperature of cold water
T_{e_1}	Equilibrium temperature after mixing of hot and cold waters during filling
T_{e_2}	Equilibrium temperature after mixing of hot and cold waters after reservoir is full
T_h	Temperature of hot water
T_o	Temperature of residual water
T_s	Temperature of steam
T_v	Temperature of vapor
T_I	Water temperature at start of eruption
T_1	Temperature at which water enters geyser reservoir
T_2	Temperature of water in just-filled reservoir
v	Velocity of jet
v_o	(1) Initial velocity of jet; (2) Velocity fluid is projected upward from geyser orifice
V	Volume
V_b	Volume of bubbles flowing up tube per unit time per unit cross section
V_o	Volume of water in reservoir after eruption
V_s	(1) Volume of steam discharged per second per unit cross section of orifice area; (2) Volume of steam formed from m_S grams of steam
V_{s_o}	Specific volume of steam
V_w	Volume of water discharged per second per unit cross section of orifice area
V_A	Volume of reservoir A
V_{G_o}	Volume of gas bubbles reduced to standard conditions
V_R	Volume of residual water in reservoir after eruption
w	Wall thickness of bubble
W	Excess heat energy

x	Fraction of total mass
α	(1) Ratio volume of spring gas to volume of spring water; (2) Deflection of the vertical
β	Constant in Na-K-Ca ratio equation
ρ	Density of liquid water
$\bar{\rho}$	Average density of superincumbent column of fluid
ρ_o	Average density of fluid (water and steam)
ρ_s	Density of steam
ρ_{s_o}	Density of steam under standard conditions
ρ_v	Density of water vapor in bubbles
ρ_w	Density of water
ρ_{w+b}	Density of water and bubbles
ρ_{w+s}	Density of water and steam
σ	Heat of vaporization of water
τ	Time
τ_e	Total eruption time of Velikan
τ_s	Length of steam phase of Velikan
τ_I	Time interval between eruptions
τ_1	(1) Time for boiling surface to move down from depth d_1 to $d_1 + \Delta d_1$, hence length of play; (2) Time to fill reservoir
τ_2	Time when channel becomes full
ψM	Radius vector between moon and earth
ψS	Radius vector between sun and earth

CHAPTER 1

Geysers of the World

1.1 Introduction

Geysers are spectacular hydrothermal events. The word geyser is derived from an old Icelandic verb, *gjose,* meaning to erupt. It refers specifically to a reservoir of hot water that intermittently and explosively ejects part or all of its contents. Activity in most geyser areas ranges over a wide spectrum: quiescent hot pools, vigorously boiling pools, dry stream jets, mud pots, and geysers (Fig. 1-1). Although there are several thousand hot springs in the world, there are not more than about 400 geysers. In Yellowstone National Park, the most extensive geyser area, the ratio of hot springs to geysers is about ten to one.

A geyser is essentially a hot spring but its unique characteristic is that it periodically becomes thermodynamically and hydrodynamically unstable. A very special set of circumstances must exist for a hot spring to erupt. It must have a source of heat. It must have a place to store water while it is heated up to just the right temperature, an opening of the optimum size out of which to throw the hot water, and underground channels adequate for bringing in fresh water after an eruption. Only very rarely does the right combination exist. When there is little water but intense heat, a steam vent called a fumarole exists. A mud pot occurs when the hot water is laden with dirt. If there is plenty of incoming water but it is comparatively cool, it is a hot pool; or if too hot, a spouter continuously spitting out steam and hot water. If the opening is too large or the reservoir so shaped that circulation can occur freely, instabilities may not be able to develop and the hot spring simply boils.

A geyser erupts when a part of its stored hot water becomes unstable, i.e., its heat content reaches some critical level of distribution. Abrupt and vigorous generation of steam occurs within the geyser comparatively close to its surface opening. The transformation of 1 g of water to steam can do as much work as the

1

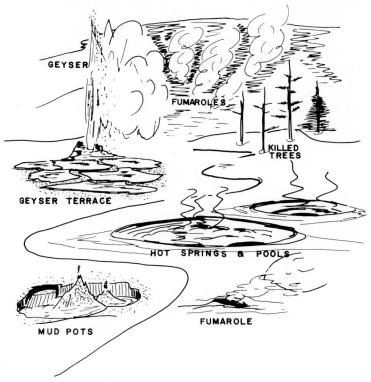

Figure 1-1. Surface features found in a typical geyser basin. (From Rinehart, 1976b.)

detonation of 1 g of explosive. Water in the form of steam occupies more than 1500 times as much volume as in the form of liquid, the same ratio as the gases generated by a solid explosive. The presence of steam greatly modifies both temperature and fluid distributions, forcibly throwing water out of the geyser and precipitating a full-fledged eruption. When the geyser has exhausted its excess heat and water, it returns to a stable condition, all set to begin a new cycle of instability. The buildup of heat usually results from hot water or steam entering the reservoir at a deep level, a few hundred to a few thousand meters below.

Although no two geysers are alike in all respects, most fall into one of two rather distinct classes and traditionally have been classed as fountain or pool, and columnar or cone geysers. Fountain or pool geysers are usually characterized by their surface pools of hot water. Their eruptions consist of series of steam and water explosions the source of which are blobs of superheated water which suddenly rise to the surface of the upper basin and flash into steam.

Columnar or cone geysers for the most part display cones or protuberances above their narrow subsurface tubes which are filled with water and are emptied partially or completely during eruptions. Their eruptions are precipitated when underlying superheated water down within the tube is heated to the point where

steam bubbles begin to form. These bubbles reduce the hydrostatic pressure of the overlying water column, more superheated water flashes into steam, and soon the whole tube empties itself catastrophically. Some columnar geysers do not build cones, their standpipes opening into the bottom of a pool formed by a raised rim of deposited sinter.

1.2 The Geyser as a Geographic and Geologic Feature

Geysers are not common geologic features. They exist only here and there in a few widely separated, highly localized regions. The most famous areas are in Yellowstone National Park in northwestern United States, Iceland, the North Island of New Zealand, Kamchatka in northeastern USSR, and Japan. There are a few isolated geysers in Chile, Mexico, Africa, the Azores, Indonesia, various Pacific islands, The People's Republic of China, the Basin and Range Province of western USA, and Unmak Island, Alaska. The USA geyser areas are located in Fig. 1-2.

There are about 200 active geysers in Yellowstone, the number varying from time to time. This is approximately 10 percent of the total number of hot springs. Of the plus or minus 3000 hot springs in Iceland, not more than 40 are active geysers. New Zealand has even fewer geysers. However it can boast of having had the largest one in the world, the now inactive Waimangu Geyser. Geyser Valley in Kamchatka contains about 100 geysers of which about 20 are as large as some in Iceland and Yellowstone. The natural geysers most widely known in Japan were those at Atami in Shizuoka Prefecture and at Onikobe in Miyagi Prefecture.

All of the principal geyser areas are found in volcanic regions and contain large quantities of rhyolite from which the geysers spout. A few emerge from basalts and andesite.

The Yellowstone hot springs are found in about 100 clusters. The locations of the several geyser basins are mapped in Fig. 1-3. These areas are found on hillsides and in valleys, most frequently close to surface water, lakes and rivers. The areas vary in size from a few square meters to a few square kilometers and can be easily identified by their mounds and terraces of light gray or buff deposits of loose sandy material, consolidated siliceous sinter, or travertine.

In the power and beauty of its geysers, Upper Geyser Basin surpasses any other geyser area in the world (Fig. 1-4). It extends 3 km along the 2 km wide Firehole River valley. Steep pine-forested, low mountains rise from both sides of the valley. Along the banks of the swift 20 m wide river is a wide, barren white strip of sinter deposited through the ages by the heavily mineralized waters flowing from the hot springs and geysers. In cold and damp weather, the strip is filled with columns of steam rising from the many hot springs (Fig. 1-5).

About 70 active geysers are located in the Upper Geyser Basin. Unusually noteworthy are Old Faithful, Grand, Giant, Castle, Giantess, Beehive, Riverside, Daisy, Artemisia, Grotto, and Lion. Old Faithful, Grand, Giant, and Bee-

Figure 1-2. Locations of geysers in Western United States.

hive play to heights of 30 to 60 m. Each geyser is surrounded by its own characteristic heavy sinter deposit. Some geysers like Grand, Daisy, and Artemisia play from pools that are nearly at ground level; others have built up their individual cones through which they erupt. Regular truncated cones have formed around Beehive and Lion geysers, whereas the Riverside cone and the especially prominent one of Castle are quite irregular. The structures at Giant and Grotto are fantastic (Fig. 1-6).

A few small geysers and several hot pools occupy Black Sands Basin located 3 km west of the Upper Basin.

Midway Geyser Basin also lies along the Firehole River approximately 10 km downriver. It contains many significant springs but only a few small active geysers. The dominant feature of the Basin is the now extinct Excelsior Geyser which was very active during the latter part of the nineteenth century.

The Lower Basin, 5 km further down the river, is the most extensive. It occupies approximately 40 km² of nearly level mixed meadow and forest land. No other basin equals it in the discharge of hot water, about 750 liters/s. Its over

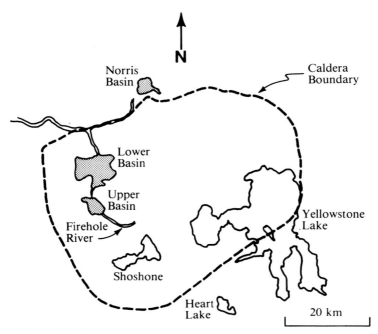

Figure 1-3. Locations of geyser basins in Yellowstone National Park.

600 hot springs occur in a dozen or so groups in the center and along river tributaries around the borders of the basin. There are 40 to 50 active geysers.

The major geysers of the Great Fountain group lie along the hot spring-fed White and Tangled Creeks. White Dome's high sinter cone is one of the largest and most picturesque in the Park. Great Fountain is the most interesting geyser in Lower Basin. It is unique in the extent, symmetry, and beauty of its sinter terraces. An irregular circular platform 50 m in diameter and up to 50 cm high surrounds the crater (Fig. 1-7). Its edges are scalloped and highly decorated with irregular humps and depressions and the several ringlike depressions trap and hold water. The crater, within a slightly raised rim, is filled with clear superheated water, often boiling vigorously around the edges. Eruptions are preceded by a period of substantial overflow and come at intervals of about 8 hr. They consist of a series of spectacular fanlike bursts of steam and hot water, exceeding 30 m, and continue on and off for about an hour.

About 1 km from Great Fountain and White Dome is a group of six geysers, three of which, Pink Cone, Narcissus, and Bead play to heights of 5 m. The small truncated cone from which Pink Cone discharges its thin jet is distinguished by its lovely color, attributed to the presence of small quantities of manganese oxide.

The Fountain group lies on a prominent 300 m² terrace that rises about 5 m above an extensive sinter plain. The group contains many hot pools, several mud

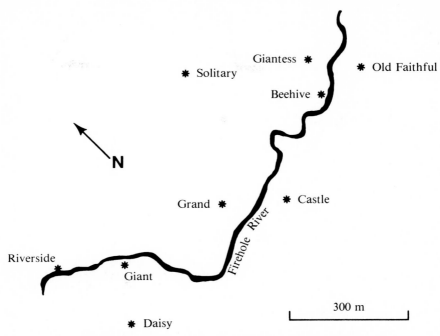

Figure 1-4. Map of Upper Geyser Basin, Yellowstone.

pots, and 11 true geysers. Fountain Geyser, next in size to Great Fountain, plays to a height of 15 m for 15 min.

The Imperial Group, located about 2 km inland from the west bank of the Firehole River contains one of the most famous of Yellowstone geysers. For two years in the late 1920s, the broad and powerful eruptions of Imperial made it one of the most spectacular in the Park.

Several of the other small groups of springs occasionally but not always include a geyser.

The Norris Geyser Basin is a 15 by 1.5 km quadrilateral area. A high pine-clad rhyolite ridge bounds it on the east, draining into shallow depressions. To the west it extends to the Gibbon River. Viewed from the ridge, the Porcelain Basin section displays brilliant varied color, blue hot pools, white sinter sheets, thin yellow sulfur incrustations, splashes of red and orange sulfides of arsenic and iron, and emerald green organic growth.

The basin is hotter than either the Upper or Lower Basins and much less wet. Temperatures of 136° to 138°C have been observed and it is the only area in Yellowstone where superheated steam vents and geysers exist side by side.

There are about 40 active geysers in the basin. Most are small, but several play to heights of from 5 to 30 m. But the world's largest active geyser, Steamboat, which plays to a height of 100 m, is there. Most of the geysers have no large associated sinter deposits and issue simply from cracks in the rhyolite rock.

Figure 1-5. Biscuit Basin, Yellowstone, on a cold morning.

This suggests that the Norris Basin is geologically young compared with Upper, Midway, and Lower Basins. Indeed, Norris, much more so than these other Basins, has undergone many rapid observable fluctuations in activity during the past century.

The stupendous Steamboat is believed to be only 100 years old, having first exploded violently out of a hillside in 1878 to form what became known as New Crater. Now it consists of two roaring steam holes still looking as though they had just burst through the surface.

En route to Shoshone Geyser Basin is the isolated Lone Star Geyser. The Basin lies further along, about 10 km south of the Upper Geyser Basin. It exhibits practically all of the features of the larger basins except in extent, 100 by 500 m, and power of the geysers. There are six geysers that play to a height of 3 m or more: Union, Minute Man, Lion, Little Bulgar, Little Giant, and one unnamed. Union is by far the most powerful, its jet reaching a height of 35 m.

Along the shores of the high mountain Heart Lake and extending up Witch Creek, about 25 km east of Shoshone, there are five separate groups of springs:

Figure 1-6. Sinter cone of Giant Geyser, Yellowstone.

Rustic, Lower, Middle, Fissure, and Upper Group. All of them contain some geysers, their eruptions varying in height from a few centimeters to 15 m.

The largest geysers in Iceland are grouped together near the Great Geysir at an altitude of 110 m above sea level about 70 km northeast of Reykjavik (Fig. 1-8). The nearly flat basin, Haukadalur (Fig. 1-9), roughly 100 by 500 m, parallels the natural fault system. There are a large number of hot springs, only a few of which are geysers. The entire area is very colorful and attractive. The two active are Strokkur Geyser (the Churn) and Smid Geyser (the Cooker) but they are but small imitators of the Great Geysir, a great performer when it erupted. Strokkur erupts to a height of about 20 m on a 10 to 15 min schedule; and Smid only becomes a geyser when soaped. The central plateau with its bordering high and extremely wet mountains lies both to the east and west of the basin.

There are two small geysers, Uxahver and Ystihver, lying side by side in the north of Iceland near the thermally very active region near Lake Myvatn. There are also a few other individual geysers scattered throughout Iceland.

New Zealand's geysers, steam vents, and boiling springs are confined largely to the Taupo Volcanic Zone, 60 by 200 km. The zone runs athwart North Island from the active Ruapehu volcano in Tongariro National Park, northeast to the also active White Island volcano lying 50 km off the Bay of Plenty coast (Fig. 1-10). This volcanically active region is strewn with conical and dome shaped volcanoes. In some regions, their eruptions have mantled the countryside with volcanic ash to form many natural lakes. In other regions, more violent eruptions

Figure 1-7. Great Fountain Geyser, Yellowstone, during quiet steam phase following an eruption.

laid down thick, flat-lying sheets of both soft and strongly welded rock called ignimbrite, often covering a few thousand square kilometers.

Geysers were formerly found within this zone at Wairakei, now the location of a geothermal electric power plant that obliterated all activity. The hilly, heavily wooded area lies along the Waikato River between 5 and 10 km downstream from Lake Taupo. Originally geysers and hot springs discharged copious amounts of hot water and steam into the river along a few kilometer stretch.

Orakei-Korako is also located on the Waikato River a few kilometers downstream from Wairakei. Before being flooded by a hydroelectric project in 1961, it was a major geyser area extending along both banks of the river for about 2 km. There was one large whitish mass of siliceous deposit called Papa Kohetu (Flat Stone) 120 m long and of equal breadth, out of which several geysers and many springs issued. The spring that supplied most of the silica to form the stone

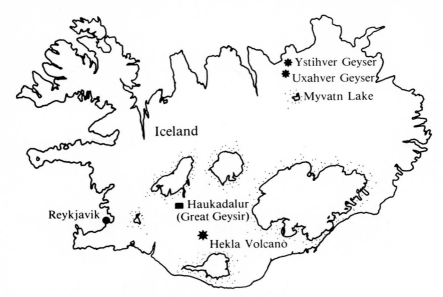

Figure 1-8. Locations of Icelandic geysers.

was a powerful fountain continually bubbling 98°C water to a height of about
1 m. After a large earthquake in 1848, it became a geyser spouting to a height of
30 m and remained so for about 2 yr.

The inactive Waimangu (Black Water) Geyser is located a few kilometers to
the southeast of Rotorua on the floor of a narrow, steep-sided valley of quite
recent volcanic origin close to Mount Tarawera, a volcano that erupted with great
violence in 1886 and altered the face of the whole area. Waimangu's eruption
resembled more that of a great mud volcano than that of the graceful water-steam
eruptions of most other geysers.

Whakarewarewa is New Zealand's most famous geothermal area. Its spectacular geysers, boiling springs, and mud pots are located just south of the city
limits of Rotorua about 45 km northeast of Wairakei (Fig. 1-11). There are more
than 500 springs in all stages of activity strung along 3 km of Puarenga Creek
where it empties into Lake Rotorua. The main geysers issue from the Te Puia
Fault where it crosses the 6000 m² hot area known as Geyser Flat and lie within a
radius of 100 m. The main activity is concentrated along the east bank of the
stream which has deeply eroded Geyser Flat. Further faulting, fissures, and
collapse pits resulting from geothermal decomposition provide passages for the
escaping steam. Most of the ground is barren, either altered to brightly colored
clays, or encrusted with sinter deposited as terraces around the springs. Seven or
eight of the springs are periodic geysers of which Pohutu is the most active. Its
heavily fissured and fractured basin is 4 m wide and extensive masses of sinter
are piled up more than 7 m.

Kamchatka is situated far to the north in the USSR (Fig. 1-12). The Kam-

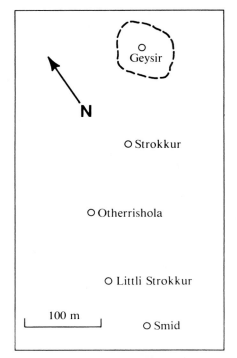

Figure 1-9. Haukadalur Geyser Basin, Iceland.

chatka geysers are all located in a 2.5 km long part of the Geyser Valley at the point where the Geysernaya empties into the Shumnaya River (Fig. 1-13). There are roughly 100 geysers altogether, only 20 of which are comparable in size and eruptive force to the Yellowstone, Icelandic, and New Zealand geysers.

Velikan (the Giant), the largest, hurls water 40 m into the air and steam to heights of several hundred meters. Each geyser is surrounded by beautiful formations of siliceous geyserite, some quite extensive such as the 900 m^2 platform on which Velikan sits. Color and form vary depending on the chemical composition of the water and the presence of thermophilic algae and various other microorganisms. Many multicolored mud pots cling to the higher hillsides.

There are about 6000 separate thermal springs associated with Japan's numerous volcanoes. Until 1924 when Atami Geyser ceased its activity there were only three well-known natural geysers in Japan: Atami, located close to the seacoast about 160 km southeast of Tokyo; and Megama and Ogama located in the mountains near Onikobe, about 80 km northwest of Sendai. All are now quiescent. These geysers issued from andesite and in general did not form any distinguishing deposits. During the last 30 years, many artificial geysers have been obtained by boring in various hot spring areas, especially near Onikobe.

The Beowawe geysers are located about 2.5 km from the village of Beowawe in central Nevada, USA. Until being drilled for possible use as a geothermal

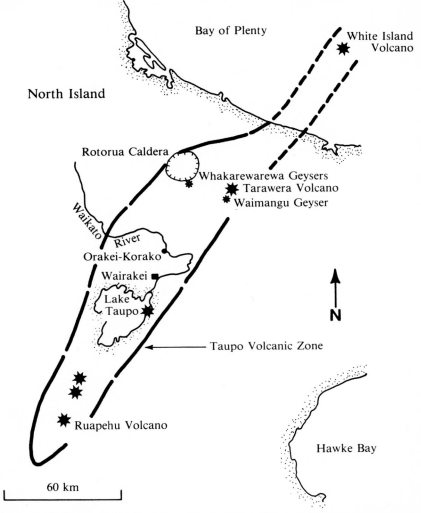

Figure 1-10. Taupo Volcanic Zone, North Island, New Zealand, showing major geologic and geothermal features. (Adapted from Healy, 1964.)

power supply, thermal activity was largely restricted to the surface of a 60 m high sinter terrace about 1 km long and 30 m wide. About 100 fumaroles, at least an equal number of hot springs, and a few geysers were scattered along the top of the terrace. Of the three geysers that erupted to any considerable degree, two erupted to a height of 1 m and the third to 4 m. A few formerly pulsating pools lying at the foot of the terrace changed into small geysers subsequent to the drilling operations. Sinter mounds, some more than 1 m high, surround the now dead geysers.

Steamboat Springs is a group of hot springs, a large number of steam vents

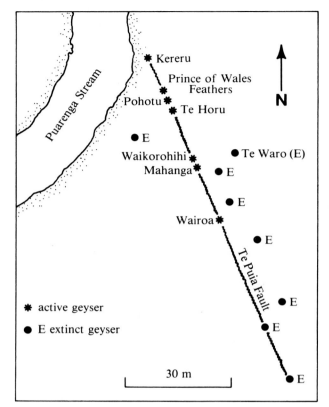

Figure 1-11. Whakarewarewa Geyser Basin, New Zealand. (Adapted from Lloyd, 1965.)

and fumaroles, and numerous small effluent seeps that issue from fissures in sinter in a hilly region near Reno, Nevada. The 20 or so geysers were small and inconspicuous before they were tampered with, although occasionally one would erupt to a height of 8 to 15 m. Now none are active.

1.3 Geyser Studies

The geysers of Iceland, especially the Great Geysir, were well known to the outside world long before those of Yellowstone and New Zealand. It was a favorite place for gentlemen travelers and scientists to visit and study. Lord Mackenzie toured Iceland extensively in 1811. R. Bunsen, the famous German chemist, best known perhaps for his invention of the common laboratory gas burner, and his assistant A. Descloiseuz made a trip to Iceland in 1847, probed the geysers, especially the Great Geysir, with thermometers and analyzed their waters for mineral content.

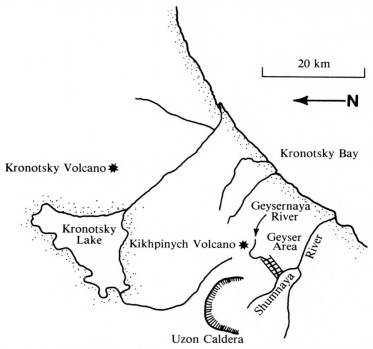

Figure 1-12. Map locating Geyser Valley, Kamchatka, USSR. (Adapted from Steinberg et al., 1978.)

Many of the original papers on thermal activity in Iceland were written in Icelandic or Danish making them nearly inaccessible, partly because of limited distribution but mostly because of language. A notable exception is the results of T. Thorkelsson's extensive studies of the hot spring and geyser activity and his discussion of his theories of geyser action.

By far the most complete description of Iceland's thermal features is given in the voluminous treatise by T. F. W. Barth who made field trips to Iceland in the summers of 1934 and 1935. He made detailed personal observations and measurements which were later supplemented by library research and consultation with knowledgeable persons.

Individual scientists, both Icelandic and foreign, have made limited studies of the Icelandic geysers from time to time: F. Birch and associates, temperature distributions within the Great Geysir vis-á-vis its eruption pattern; and Rinehart, seismic signatures of several of the geysers.

The Icelandic government has not imposed especially stringent restrictions on the manipulations of the geysers and consequently some rather interesting experiments have been carried out in an effort to stimulate and revive geyser activity.

Ferdinand von Hochstetter visited New Zealand toward the end of 1858, traveling at a leisurely pace through the main thermal belt from Lake Taupo to

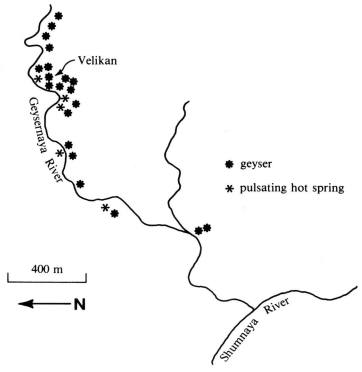

Figure 1-13. Geyser Valley, Kamchatka. (Adapted from Steinberg et al., 1978.)

Rotorua. He was a keen observer and after his return to Germany published a book incorporating his observations on the wonderful hot springs and geysers. The book has been a great help in assessing the substantial changes occurring since the 1850s, especially conditions pre and post Tarawera's violent 1886 volcanic eruption so close to the important thermal areas.

There is now a small Maori village within the Whakarewarewa thermal area but before 1886, few Maoris lived there permanently. Hochstetter saw only an insignificant settlement there. Almost certainly the actual geyser area had been occupied on several occasions before when large numbers of Maoris began to settle in the nearby Rotorua region in the fifteenth century. Part of Whakarewarewa was acquired by the New Zealand government in 1898 and made a scenic reserve. The remainder of the area and Whakarewarewa Village are still owned by the Maoris. The geyser area was administered by the Department of Tourist and Health Resorts until 1962 when the Maori Arts and Crafts Institute assumed its administration.

In recent years, scientists of the New Zealand Geological Survey, especially E. F. Lloyd, have studied the geysers carefully.

Yellowstone's spectacular features were not appreciated by a wide audience

until around 1870. The vivid descriptions brought back for several decades by normally tall-tale telling explorers and trappers were simply not believed. Finally, in the late 1860s and early 1870s, several expeditions staffed partly by military engineers and partly by civilian scientists confirmed the beauty and uniqueness of the numerous and fabulous thermal features of the area.

The Cook-Folsom-Peterson civilian expedition that went in 1869 saw it all, but then were understandably reluctant to report the wonders for fear of unbelieving laughter. Eventually they wrote about their travels and earlier maps were improved.

The 1870 Washburn-Doane-Langford civilian party accompanied by a military escort did, however, fully disclose the nature of their findings. N. P. Langford, the diaryist of the group, wrote complete accounts in the popular magazines of the day and lectured around the country. Camping in the Upper Basin along the western bank of the Firehole River midway between Castle and Beehive geysers, the group was extremely lucky. In the 22 hours they were there, seven of the largest geysers erupted. Many bear the names the party gave them—Old Faithful, Beehive, Grand, Giant, Giantess, Fan, Grotto, Castle.

The next year, 1871, the U.S. Geological and Geographical Survey sent the Hayden Expedition in specifically to explore and photograph the scientific wonders of the area. W. H. Jackson, the famous pioneer photographer of the western frontier, accompanied it. The reports and photographs coming out of this and the subsequent expeditions of 1872 and 1878 were extremely helpful in the steps that led to Yellowstone becoming in 1878 the first National Park, an earlier idea revived by the Washburn party.

Beginning with A. C. Peale, the first Superintendant of the Park, and continuing to the present, regular observations have been made on the activity of the thermal springs and geysers, their temperatures, and other matters including seismic activity. For the most part, these observations are reported in monthly and annual internal reports and special technical notes when the occasion warrants. A. Hague, a geologist with the Survey, from 1883 to 1893 began a systematic investigation of the hot springs and geysers. His observations were supplemented in 1896, 1897, 1902, 1911, and 1915. Much of the information was reported in his Annual Reports.

E. T. Allen and A. L. Day, geologists of the Geophysical Laboratory of the Carnegie Institution of Washington, spent several months of each year from 1925 through 1932, with the exception of 1931, in the field at Yellowstone making observations on the hot springs and geysers, supplemented in the laboratory with thorough chemical and microscopic examination of specimens. The results of their studies are by far the most exhaustive description available on the general features of the Yellowstone thermal area.

G. Marler was employed as a summer ranger starting in the late 1930s and continued until his retirement in the early 1970s as a full-time geyser and hot springs watcher. His many reports thoroughly chronicle the thermal activity in the Park during this time. Much of this work is now being continued by the ranger-geologist R. W. Hutchinson. In addition, the rangers stationed at the Old

Faithful Information Center have monitored the Upper Geyser Basin closely during the summer since 1921, recording observed and reported geyser activity in the Basin, especially every daylight eruption of Old Faithful. A recent extensive guide book by T. S. Bryan has updated and enlarged the Park's information on its geysers.

The U.S. Geological Survey has taken a growing and keen interest in the Yellowstone area partly because of the emergent interest in geothermal energy, partly because of the relevance of its geology to space exploration, and partly because of scientific curiosity as to the physical and chemical processes involved in geyser and hot spring activity. The studies of D. E. White, R. O. Fournier, L. J. P. Muffler, and A. H. Truesdell have involved chemical and isotopic analysis of many hot springs and geyser waters, and temperature and pressure measurements as a function of depth in several specially drilled holes a few hundred meters deep. Other teams of geologists and geophysicists have carried out seismic, magnetic, and heat flow studies, all of which have resulted in a better understanding of the geological, thermal, and hydrologic nature of Yellowstone.

Individual scientists and groups of scientists have also worked in the Park from time to time. For several years in the mid 1960s, V. C. Schaeffer, under National Science Foundation auspices, brought small groups of senior scientists of diverse disciplines to the Park under the severe climatic conditions of midwinter. They carried out studies of their own choosing on hot spring and geyser activity and environmental effects. Also, the team of F. Birch, I. C. Graton, and G. C. Kennedy studied the temperature regime in geysers, especially Old Faithful, as a function of depth and time; F. D. Bloss and T. F. W. Barth compared geyser theories with observed eruption patterns; T. D. and M. L. Brock investigated thermophilic organisms found in and around the hot waters; R. B. Smith and his collaborators delineated earthquake activity; and the author was primarily interested in establishing causes of periodic, episodic, and secular changes in geyser activity.

The Beowawe geysers were first noted by a U.S. Army cavalry patrol in 1867 and the published account by A. S. Evans appeared not much later. Two geologists, T. B. Nolan and G. H. Anderson, stopped there in 1932 and detailed the geology of the area and its associated thermal activity. Seismic disturbances generated by the geysers and hot springs were recorded and published about forty years later by Rinehart.

For many years D. E. White and his associates have studied the Steamboat Springs area in Nevada. Manipulating the natural geysers and the geysering wells and using this information they developed many fruitful concepts and theories concerning the action of geysers.

There has been no freedom of access of foreign scientists to the Kamchatka region in the USSR since the geysers were discovered there in 1941. However, the Institute of Volcanology at Petropavlosk has carried along a vigorous hydrothermal and volcanic program of investigation and many of their findings have been published and translated along with photographs and maps. I. F. Golovina and N. N. Malov presented a rather complete theory of geyser action based on

their observations of Velikan. G. S. Steinberg and his associates have been extremely active both in the field where they have manipulated one of the geysers to learn more about it, and in the laboratory where they have used model geysers to confirm their geyser theories.

A number of productive studies have been carried out in the field in Japan. K. Honda and T. Terada established the eruption pattern of the now dormant Atami Geyser and were among the first to note and attempt to explain why some geysers erupt on a bimodal schedule. Later K. Honda and T. Sone presented a fairly complete description of the mineral springs of Japan. Y. Nomura, K. Noguchi, and somewhat later, I. Iwasaki investigated thoroughly the geochemistry of the waters of both the natural geysers and the geysering drilled wells. H. Nogoshi and Y. Motoya recorded, analysed, and attempted to account for the origins of the seismic tremors generated by each of the six drilled geysers at Onikobe.

1.4 Behavioral Characteristics of Some Geysers

No geyser looks or acts the same as any other. Each has its own arrangement of reservoirs and tubes, water supply, and heat source. However, by closely observing the activity of individual geysers and groups of them, it is possible to learn much concerning the general nature of operational modes.

A deep pool marks the location of the Great Geysir, sitting atop a large flat mound. The geyser, which was quite active during the nineteenth and early part of the twentieth centuries, became dormant in 1915. It was rejuvenated in 1935 by cutting a narrow gate in the rim of the basin, but now it is again inactive.

One of the earliest of the scientist-travelers to Iceland, Krug von Nidda, describes an eruption of the Great Geysir (Fig. 1-14):

> A thick pillar of vapor rose to the clouds with the rapidity of an arrow, and included in its interior a column of water which rose with a wavering movement from the mouth of the geyser to a height of 80 or 90 feet in the air, but which soon fell to half the altitude. Detached smaller jets ascended much higher and others were projected in inclined arcs from the envelope of the vapor. Soon the vapory clouds were dispersed and displayed the column of water, which, separated into innumerable jets, was projected in a straight line upwards, then spread itself out at its summit like a pine tree, and afterwards descended in the form of a fine dust-like rain. The clouds of vapor speedily collected around their nucleus, in order to exhibit it anew in a still more surprising form. Several times the gigantic power seemed worn out, and the column suddenly disappeared; but the earth was anew agitated—dull thunder rolled beneath and the column of water was elevated into the air by the steam with renewed force. The activity of the spring lasted including the short periods of repose, about ten minutes. The column of water then sunk back into the mouth of the geyser, and stillness was restored. The water fell to a considerable depth in the pipe, and then began to rise but slowly.

Waimangu, the largest geyser ever observed to erupt, was very active from January 1900 to November 1904. Its dormant site is now almost unidentifiable by

Figure 1-14. Old engraving showing The Great Geysir in full eruption.

the growth of vegetation. When active, an eruption, occurring at about 36 hr intervals, threw jets of mud, rocks, water, and steam to heights of up to 450 m in one large explosive burst. An eruption came on quickly and almost unheralded. Normally its 100 m diameter pool would be quiet except for small bubbles rising to the surface and steam drifting lazily across it. Suddenly the pool would begin to seethe and boil violently and then almost immediately explode. The black water, mud, and rocks hurled into the air would soon be whited out by billows of steam. Within a few minutes the pool regained its composure, leaving no sign of its violent action except for the rubble strewn about.

Excelsior's now quiet pool was during the 1880s the most powerful geyser ever seen in Yellowstone, erupting explosively to heights exceeding 100 m and throwing out huge quantities of water and numerous rocks. Its eruptions came at 60 to 80 min intervals, each one lasting 5 to 15 min. One observer saw 63

eruptions during a 10 day period. The prodigious amount of discharged water doubled the flow of the Firehole River. The extinction of Excelsior in 1888 has been ascribed to the enlargement of its opening by the ejection of rocks or possible damage within its reservoirs. The huge 60 by 100 m rectangular crater left is now filled to within 5 m of its top with steaming hot, clear light-blue water which flows out in large quantities and rushes down a steep sinter incline into the river. Bright red and yellow colonies of hot-water algae flourish along the edges of the cascading water. The rate of discharge of water from the spring is very high, about 60 liters/s.

Old Faithful is perhaps the best known in the world (Fig. 1-15). Its 4 m high, 50 by 70 m sinter mound holds a commanding position at one end of the Upper Geyser Basin. The opening of the geyser, 1.5 by 3 m, is essentially a widened

Figure 1-15. Old Faithful Geyser in full eruption.

segment of a fracture that extends across the top of the mound. There is no surface pool, the water being ejected directly out of the opening.

An eruption is heralded by premonitory splashes that rise to a height of a few meters. The eruption starts with a higher splash, quickly followed by another and another, each noisily rising to a greater height before the others have completely fallen to the ground. It takes a minute or two for the ebullient column of gushing water to reach its peak. One or two spurts may shoot even higher before the column begins to fall in easy stages. Total water play will last from 2 to 5 min, followed by several minutes of steam play during which steam in great quantities billows out of the opening. Old Faithful is perhaps most beautiful on a warm day when the water jet is not obscured by clouds of steam. The time to the next eruption can range from 30 to 100 min. It is predictable to within 5 min based on its just-passed length of play.

For symmetry and grace in both sinter cone and water jet, no geyser can surpass Beehive, located at the lower edge of Geyser Hill in Upper Geyser Basin. The cone, shaped like a traditional beehive, is 2 m in diameter and stands 1 m high on an otherwise smooth sinter-covered hillside (Fig. 1-16). The orifice, 70 cm at the top and quickly narrowing with depth, restricts the geyser's discharge to a narrow hose-like jet. The jet plays to a height of over 60 m for anywhere from 2 to 5 min. Some years it plays with great regularity every 8 to 12 hrs, whereas other years it is nearly dormant.

Castle Geyser, so-named because of its impressive castle-shaped cone, is also in Upper Geyser Basin (Fig. 1-17). The brilliant grayish-white cone is very large, 4 m in height and 12 m in diameter at its base, forming a cup-shaped circular basin at its top 1 m in diameter ending in a narrow vertical channel. Castle has a sustained 1 hr or more long eruption, discharging its steam and water in successive spurts some of which may reach 30 m in height. As the eruption proceeds, the proportion of water, originally high, gradually decreases until toward the end only steam is being discharged. Eruptions occur every 6 to 9 hr.

The eruptions of Grand Geyser are considered by many to be the most magnificent in Yellowstone. Grand is a fountain geyser lying on the east bank of the Firehole River. During its quiet periods, when empty, it is an unobstrusive 20 m diameter pad of sinter not much above ground level. While it is filling with water, it can easily be mistaken for an ordinary hot pool. Much of the time it erupts about three times in a 24 hr period. An eruption consists of about six to eight, but sometimes as many as 40 explosive bursts spaced a few minutes apart. Each burst, lasting as long as 2 min, shoots jets of hot water and steam to as high as 70 m. Successive bursts seem to grow in splendor. The action usually lasts 10 to 20 min but may continue for as long as an hour. The water remaining in the pool is then powerfully sucked back underground.

Steamboat Geyser is at present the largest active geyser in the world and its major eruptions are the tallest and most powerful displays in the known recorded history of the Park. Only Excelsior has exceeded it in massiveness and volume. Steamboat has been active at least since 1878 when it was first observed by Park personnel. In some years it plays rather frequently, sometimes as often as every

Figure 1-16. Old engraving of Beehive Geyser in full eruption.

two weeks but in other years, no major eruptions occur. In March, 1978, the first major eruption in nine years took place and it has erupted several times since. Between major eruptions, Steamboat is mildly active, throwing and splashing water on a fairly frequent schedule to heights of eight to 15 m. At the time of a major eruption, Steamboat hurls jets simultaneously from two neighboring vents located on the side of a hill. The jets which are inclined 45°, rise to heights ranging from 90 to 120 m. An eruption begins abruptly, apparently without any

Figure 1-17. Castle Geyser during steam phase.

noticeable warning, as a column of water that maintains a nozzle-type discharge, rising gradually and mushrooming out when it reaches its maximum height. The west vent pours forth mostly clean water and the east vent gray muddy water and rocks, ranging from sand grains to pieces eight to 10 cm. Some of the large pieces fly as far as 60 m, sand and gravel up to 150 m, and trees as far away as 100 m become covered with a fine layer of gray mud. This water phase lasts from 10 to 40 min, and is followed by a powerful steam phase lasting 30 to 40 hr. Toward the end, water jets from the geyser's openings to heights of from 10 to 15 m.

The Geologic, Thermal, and Hydrologic State of the Earth

2.1 Geologic Features of the Earth

To a first approximation, the earth is a 12,700 km diameter 5.976×10^{27} g sphere that rotates once a day on its own axis, follows an elliptical path around the sun once a year, and is circled by the moon about once every 28 days. It was formed as a more or less homogeneous mass about 4.5 billion years ago, but soon began to undergo gravitational differentiation, the lighter materials coming to the surface and the heavier ones settling down toward the center. It now has a dense liquid iron-nickel 6900 km diameter core that contains an even more dense solid 2700 km diameter inner core. The liquid core is in turn surrounded by a 2900 km thick, much less dense solid mantle that accounts for about two-thirds of the total mass of the earth. A still less dense crust varying in thickness from 5 to 70 km, overlies the mantle, the thinnest parts of which lie under the ocean. Thicker sections form the continents and the thickest parts lie beneath the mountains (see Fig. 2-1).

The heterogeneous mantle constantly interacts with the crust that it moves, and with which it mixes (Fig. 2-2). As new oceanic and continental crustal material is formed, "old" crustal material sinks back. A rather abrupt density interface between the crust and the Upper Mantle is referred to as the Mohorovicĭc discontinuity, or Moho for short, after the man who first described it.

The density of the earth ranges from about 2.7 g/cm^3 in the crust, through 5.7 g/cm^3 in the mantle, 10.2 g/cm^3 in the liquid core to 11.5 g/cm^3 in the inner solid core. Oxygen and silicon are the most abundant chemical elements along with significant amounts of aluminum, iron, potassium, and magnesium (see Table 2-1). The most common compounds are the silicates and the metallic oxides (Table 2-2).

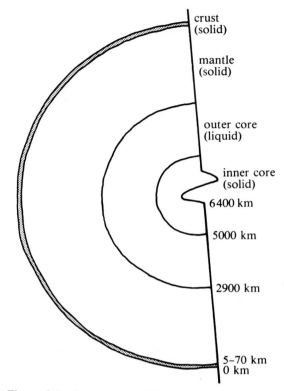

Figure 2-1. Cross section of the earth showing layering.

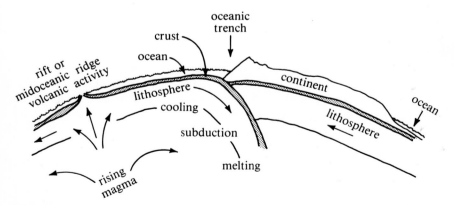

Figure 2-2. A model of the crust and mantle of the earth showing movements of the plates and lithosphere, and thermal convection of the viscuous mass beneath. (Adapted from Wahl, 1977.)

Table 2-1. Relative Abundances of Elements in the Earth

Element	wt %
Oxygen	46.6
Silicon	27.7
Aluminum	8.1
Iron	5.0
Potassium	3.6
Magnesium	2.1

Table 2-2. Most Prevalent Compounds in Rocks

Compound	wt %
Silica (SiO_2)	40 to 80
Magnesia (MgO)	0 to 45
Alumina (Al_2O_3)	0 to 25
Oxides of iron	0 to 20
Lime (CaO)	0 to 20
Soda (Na_2O)	0 to 16
Potash (K_2O)	0 to 12

Rocks are usually classified into three types: igneous, metamorphic, and sedimentary. The first are those which are produced by igneous activity such as volcanoes and transports of magma (molten rock). Most are composed of various complex combinations of aluminum, potassium, magnesium, iron, titanium, the phosphates, and containing large percentages of silicon and oxygen. Typical are those shown in Table 2-3. Some of the more common igneous rocks are the basalts, andesites, dacites, rhyolites, and diorites. The basalts which are basic, contain about 45 percent silicon dioxide (SiO_2) which increases in percentage in the more acidic rhyolites up to as much as 75 percent (see Table 2-4).

Metamorphic rocks are either igneous or sedimentary rocks that have been mineralogically transformed in situ by the application of heat, or pressure, or through chemical reactions. Some of the more common metamorphic rocks identified mostly by structure are: slate, a metamorphosed clay; and schist, gneiss, granulite and hornfels, metamorphosed granite. Others based more on minerology are: marble, recrystallized calcite or dolomite; quartzite, recrystallized quartz, amphibolite, hornblende, plagioclase rock; soapstone, mostly talc; and tactite, derived from limestone.

Sedimentary rocks and unconsolidated sediments fall roughly into two classes, clastic and chemical. The clastics are made up of debris of all kinds such as volcanic ashes and the ablated and often transported quartz, biotite, plagioclase and feldspar fragments that form sands, gravels, and clays. The chemical sedimentary rocks are formed as a result of in situ chemical reactions such as the deposition of calcium during the formation of caliche or as occurs in limestone

Table 2-3. Composition of Minerals Forming Igneous Rocks (Adapted from Verhoogen et al., 1970)

Mineral	Chemical Formula
Albite	$NaAlSi_3O_8$
Anorthite	$CaAl_2Si_2O_8$
Apatite	$Ca_5(PO_4)_3$
Diopside	$Ca(Mg,Fe)Si_2O_6$
Hematite	Fe_2O_3
Hypersthene	$(Mg,Fe)SiO_3$
Ilmenite	$FeTiO_3$
Magnetite	Fe_3O_4
Nepheline	$NaAlSiO_4$
Olivine	$(Mg,Fe)_2SiO_4$
Orthoclase	$KAlSi_3O_8$
Plagioclase	$(Na,Ca)Al(Si,Al)Si_2O_8$
Quartz	SiO_2

Table 2-4. Mineral Constituents of Common Igneous Rocks (Adapted from Verhoogen et al., 1970)

Rocks	Constituents
Volcanic	
basalts	calcic plagioclase, pyroxenes, olivine
andesites	medium plagioclase, pyroxenes, hornblende
dacites	" " " "
rhyolites	quartz, alkali feldspar, sodic plagioclase
Plutonic	
peridolites or ultramafic rocks	olivine, pyroxenes
gabbros	calcic plagioclase, pyroxenes, olivine
diorites	medium plagioclase, hornblende
granodiorites	quartz, medium or sodic plagioclase, potash, feldspar, hornblende, biotite
granites	quartz, potash, feldspar, sodic plagioclase, biotite, hornblende

and coral formations; the deposition of the skeletal remains of living creatures; or the deposition of geyserite, travertine, and serpentine around hot springs. The size and variation in size of a clastic rock determine its classification. Gravels, forming conglomerates and breccias, range in size from 2 mm in diameter on up; the sands, forming sandstone, from 1/16 to 2 mm; and the muds, forming claystone, from very fine to 1/16 mm. The chemical composition determines the classification of the other sediments (see Table 2-5).

The continents, and especially geothermal and volcanic areas, are made up largely of igneous rocks, either in their original state or highly metamorphosed.

Table 2-5. Chemical Classification of Sedimentary Rocks (Adapted from Verhoogen et al., 1970)

Rocks	Composition
Limestones*	mainly calcite, $CaCO_3$
Dolomites	mainly dolomite, $CaMg(CO_3)_2$
Evaporites	gypsum, $CaSO_4 \bullet 2H_2O$; anhydrite, $CaSO_4$; rock salt, NaCl (halite); various other carbonates, sulfates, and borates
Phosphate rock*	apatite, $Ca(PO_4)_3(F, Cl, OH)$
Ironstones	hydrated iron oxides (limonite), carbonate (siderite), and silicates of the chlorite family
Coal*	organic (carbon, oxygen, hydrogen)

* Largely or partly biogenic

Granite, the most common continental rock, forms by crystallization of a silicate melt. Consisting of quartz, feldspar, and mica, chemically it is 65 to 70 percent silicon dioxide, with the next most abundant constituent being aluminum oxide (Al_2O_3). Rhyolites and andesites, also rich in silica, and volcanic basalts are abundant in geothermal areas.

According to the theories of plate tectonics which have been refined during the past 15 years, the crust of the earth is in a constant state of turmoil. Some seven or more continent-size rigid plates are continually moving laterally with respect to each other. The plates deform and fracture as they interact, breaking into blocks, storing up in situ stresses, precipitating earthquakes, building mountains, creating valleys, and generating deep systems of fractures through which molten rock and hot gases can flow from great depths to the surface. The main features of the mechanics of these interactions are illustrated in Fig. 2-3.

For the most part, it is the separation and coming together of these plates, sea floor spreading and continental drift, which have led to the development of the volcanic areas of the world and their associated geothermal areas. Figure 2-3 also depicts locations of the principal geothermal areas usually colocated with volcanoes which have been active in historic times. Icelandic geysers lie atop the mid Atlantic ridge; those of Kamchatka, Japan, and New Zealand lie along the Ring of Fire which surrounds the Pacific. A major exception is Yellowstone whose geysers are presumed to lie above a shallowly buried hot magma body that now is the active front of a system of volcanic foci or magma plumes which has migrated progressively northeastward for 15 million years along the trace of the eastern Snake River Plain.

Much of the earth's crust is heavily fractured and fissured close to the surface, that is, blocks or pieces of the crust are separated or displaced either horizontally or vertically with respect to each other. Such separations or displacements are termed faults. They play an important role in shaping the surface features. Generally faults are most profuse in geologically active or formerly active areas: boundaries of continental plates, rift valleys, subduction zones, ocean ridges, sedimentation areas, and especially volcanic regions. These faults provide the

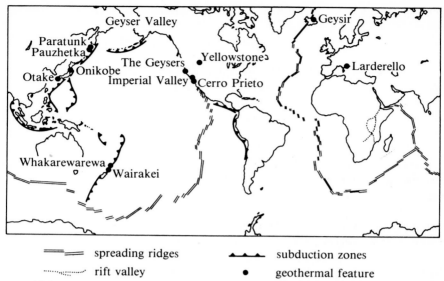

Figure 2-3. Map showing features of tectonic plates and geothermal areas.

channels through which the bulk of the flow of steam and hot water takes place in thermal areas. At Wairakei, for instance, several patches of steaming ground are localized along the surface traces of a number of northwest striking faults.

The dimensions of the fracture surfaces as well as the displacements span several orders of magnitude ranging from a few centimeters to many kilometers. Most fault surfaces are approximately planes and have a more or less constant orientation, generally lying neither exactly vertical nor exactly horizontal.

Separated fracture surfaces are ideal catch basins for alluvial fills which being porous, permit easy circulation of geothermal waters. New fault fracture surfaces exhibit bumps, curves, and corners which are partially or totally destroyed as the two sides slide and work past one another. This working often produces a finely powdered rock dust called gouge, and rock rubble called fault breccia. New fractures appear, both aligned with and athwart the fault plane, and the openings soon fill with soil and alluvium.

Faults are the response of rock masses to stresses in the earth's crust which originate from earth movements and tectonism, or thermally from cooling of igneous rocks. Erosion can uncover residual stresses formed at depth. Deposition of sediments can cause new stresses to develop. Fracturing relieves the stress. It may occur catastrophically as during an earthquake, intermittently but noncatastrophically, and continuously.

Usually the distribution of stress responsible for the pattern of fractures can be inferred. Thus radial and ring faults, such as are found in Yellowstone, Kamchatka, and New Zealand, occur around volcanoes and calderas where the surface rock is first uplifted and then dropped. The straight faults in the spreading rift valleys are tensional features lying parallel to the valley. Creation of a fault or

movement along a major fault system greatly modifies the distribution of stress, often causing secondary systems of fracture to develop. Short east-west transform faults found along the mid Atlantic ridge in Iceland are an example of such a secondary system.

The principal sources of information about faulting are surface or near-surface geologic mapping, mining operations, drilling, and geophysical probing, using the wide variety of methods now available. Recent satellite photography has been especially fruitful in delineating some of the larger-scaled fault patterns. Active faults, of course, are easily located from the earthquakes that occur along them.

Rift valley faulting and caldera faulting are the two systems of most relevance to the circulation of geothermal fluids. A rift valley is an elongated valley formed by the depression of a block of the earth's crust, called a graben, between approximately parallel faults or fault zones (Fig. 2-4). The block sinks as the sides pull apart. Caldera faulting is responsible for the major geyser areas. A caldera is often left in the aftermath of a volcano. It is a more or less circular depression with a diameter many times greater than that of the included volcanic vent. Either an explosion or a collapse or both would generate the pattern of ring and radial fractures. However created, they are underlain by a diverse and complex assemblage of rocks comprised of dikes, sills, stocks, and vent breccias; crater fills of lava; talus beds of tuff, cinder, and agglomerate; fault gouge and fault breccias; talus fans along fault escarpments; cinder cones; and other volcanic products such as rhyolites and obsidian. The structure and fracture patterns around the Timber Mountain, Nevada, caldera have been well mapped and are shown in Fig. 2-5.

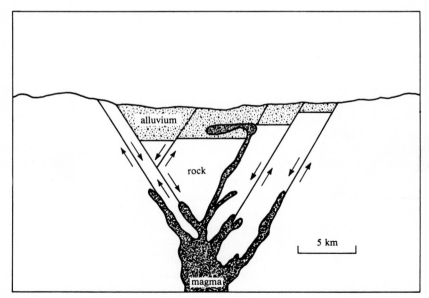

Figure 2-4. Inferred geologic section of Dixie Valley, a rift basin in Nevada. (Adapted from Thompson and Burke, 1974.)

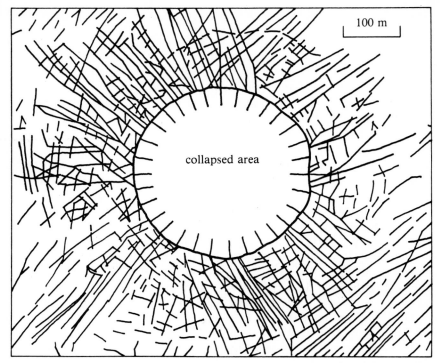

Figure 2-5. Plan view of Timber Mountain Caldera, Nevada, showing distribution of faults. (Adapted from Carr and Quinlivan, 1966.)

Most of Yellowstone Park lies within a huge 70 km long by about 45 km wide collapsed volcanic caldera that formed suddenly and violently about 600,000 years ago after a series of milder episodic extrusions of rhyolitic lavas extending over a period of 1.5 million years. This explosive eruption threw out more than 900 km^3 of rhyolite, pumice, and ash. Hot molten lava continued flowing intermittently until about 60,000 years ago. Then the glaciers moved down from the north. When they melted, the gravels they had been pushing along filled the low places, insulated the hot lava, and kept it from cooling rapidly. The Park now is a mosaic of about 20 large, very wet basins, each fringed by mountains, which are extremely efficient in collecting and storing the water that falls as heavy rains and snows. Geologic and geophysical evidence has been collected recently which fairly definitely establishes the existence of a large, shallowly buried magma body underlying the whole of Yellowstone.

Except for Norris Geyser Basin and Mammoth Hot Springs, the major thermal areas of Yellowstone occupy areas in the deep seated ring fracture zone of the Yellowstone caldera which presumably is located directly above the thermal area's main source of heat, the hot magma body lying about 5 km below the caldera and extending a few kilometers beyond its edge. Norris Geyser Basin

most certainly lies along a fracture extending radially outward from the main ring fracture.

The small area at Beowawe where geysers are found is on the southwestern edge of Whirlwind Valley, a northeasterly trending valley located in Nye County, central Nevada. The valley is one of the basins of the Basin and Range geologic province that covers large areas in Nevada, Arizona, New Mexico, eastern California, Utah, eastern Idaho, western Montana, and southern Oregon. These basins of which Dixie Valley has been by far the most studied (Fig. 2-4), are formed by spreading of the earth's crust. Many deep-seated tensional fractures and faults are generated in the process providing channels for hot magma to come up from below and water to drain down from the top, a perfect setup for the formation of a hydrothermal area. Indeed, there are many hydrothermal areas in the Basin and Range Province but geysers seem to have developed only at Beowawe and Steamboat Springs.

The alluvium-filled valley at Beowawe is wide and flat-floored with the ground water level near the surface. The rock ridges bordering the valley are fine-grained diabasic basalt, containing the minerals conducive to the formation of the plumbing required to make a geyser function. There is as well an abundant water supply to keep the geyser waters alkaline. The geyser area occupies a narrow belt at the contact between the alluvium and the lava ridges, apparently the site of a recent fault. There is indication that movement continued along the fault after the deposition of sinter had already begun.

Steamboat Springs lies on the northeast edge of the Steamboat Hills, a 23 km^2 bedrock mass. About 20 km south of Reno and 10 km west of Virginia City, Nevada, it also lies in one of the basins of the Basin and Range Province. This structural trough is about 10 km wide and lies between the Carson Range of the Sierra Nevada on the west and the Virginia Range on the east. The springs are structurally related to recent faults, the active ones lying along a 1560 km long area.

The younger rocks are largely volcanic, consisting of Tertiary andesite, late Tertiary or early Pleistocene basalt, and Pleistocene rhyolite. The heat source is probably a recent volcanic dome of pumiceous rhyolite pushed up from below.

The hot springs of the volcanically active Taupo Zone on New Zealand's North Island stretch along three parallel northeast striking lines. One line connects the two volcanoes, Tongario and Whakari. The hot springs of Lake Taupo, the fumaroles of Kakaramea mountain and the hot springs around Rotomahana lie along this line. The Puias of Orakei-Korako and of the Pairoa range form the second line; and the hot springs and geysers at Rotorua and the solfataras of the Rotoiti, the third line.

Whakarewarewa lies in the ring fracture zone of the large, 18 km diameter Rotorua caldera. Pohaturoa, an elongated precipitous rhyolite dome extruded from a fault as viscous lava dominates the area. Other faults have been defined; some are inferred from hot spring and hot ground alignments. Pohaturoa, Whakarewarewa, and Puarenga faults which trend northeast, nearly parallel to the edge of Rotorua caldera and the crest of Pohaturoa dome, were associated

either with formation of the caldera or of the dome. Either way, they probably go quite deep. Other faults, including the Te Puia Fault, which trend north-northwest are younger.

A quick inspection of the Icelandic topography is a convincing demonstration of the fact that a close connection exists between hot springs and volcanic activity. There the volcanoes, having issued on fissures, form a system of parallel lines, whose north-northeastern extremity corresponds perfectly with the parallel spreading of the principal valleys and elevated ridges, and the numerous volcanic fissures and dikes. The hot springs and fumaroles are similarly aligned (Fig. 1-9).

Kamchatka has been described by the Russians as "the end of the world, the land of the sunrise, a land of wonders and surprises, of fire breathing volcanoes, endless steppes, dense forests, broad running rivers, and mountains." It is not surprising to find that the heat of the many active volcanoes and the wetness of the rains and snows have combined to spawn a field of geysers in a valley along the Geysernaya River (Fig. 1-13).

The Pacific Ring of Fire goes directly through Japan which has many volcanoes and hot springs. Only a very few of the springs, however, become geysers. Some artificial geysers have been formed by drilling. A logical but not necessarily correct explanation of this apparent contradiction is that for the most part the springs issue from andesite, a rock less conducive than rhyolite to the formation of the structure required for a spring to evolve into a geyser.

2.2 The Earth as a Source of Heat

Geoscientists are still far from a complete understanding of the geothermal regimen of the earth, but on the basis of all the evidence now available, the following rather general conclusions can be drawn:

- formation of the earth and its separation into core, mantle, and crust must have been rather complicated events and are not understood in detail;
- the earth is not, and presumably never has been, in a thermal steady state; and
- thermal conditions in the upper few hundred kilometers on the average have not changed very much in the past 2 billion years, except locally.

The earth is certainly a tremendous storehouse of energy. Several thousand measurements made all over the earth indicate that on the average, 1.2 μcal/s flow through each square centimeter of surface. In the geyser areas, heat flow is, of course, many times higher. At Yellowstone the average is 20 to 30 μcal/s/ cm^2. Possibly such heat flows have been going on since the earth was formed. Since the surface area of the earth is 5.1×10^{18}cm^2, and there are 3.15×10^7 seconds in a year, the total heat flow is reckoned at 2.4×10^{20} cal/yr, or in work units, 10^{28} ergs/yr. Assuming that the heat flow has remained more or less constant during the last billion years, the total flow over that period through each

square centimeter of surface would have been 4×10^{10} cal. If this heat had come from the combustion of coal, which it obviously did not, it would have been necessary to burn a layer of coal 20 km thick under each square centimeter of the earth's surface.

Heat flow by conduction up through the surface of the earth is not the only manner in which the earth is dissipating heat and mechanical energy. The continual building and destruction of mountains takes 10^{24} ergs/yr; earthquakes release 10^{26} ergs/yr or 26×10^{10} kw-hr/yr; and the 800 volcanoes of the world produce about 1 km^3 or 3×10^{15} g of lava per year, releasing 1.2 cal/yr of heat. The energy of one moderate volcanic eruption is equal to 10^{22} ergs.

The heat stored in the outer 100 km of the earth at a temperature higher than that of the surface is about 2×10^{28} cal. This amount of energy is equivalent to the heat lost by conduction at the surface of the earth for nearly 100 million years, to solar radiation received by the earth for 10,000 years, to 2×10^{22} kw-hr, or to the heat content of 3×10^{18} tons of coal.

As far as heat sources are concerned, the heat could simply be that remaining from an originally hot earth following its formation. As the original temperature of the earth is unknown, it is impossible to estimate how much heat is left. Lord Kelvin showed many years ago that even if the earth were originally molten, the original heat dissipated by conduction through the rocks to the surface would only be a small part of the fraction of that now observed.

The earth probably formed originally as a mass of uniform density, later separating into its now very dense core and less dense mantle and crust. This separation represents a decrease in potential energy of the order of 10^{38} ergs, energy that would be released in the form of heat whether the separation took place rapidly or slowly.

The gravitational pull of the sun and the moon on the earth could account for about 10 percent of the heat, a rather significant amount of the total. The pull deforms the solid part of the earth as well as causing the ebb and flow of the ocean tides. While the strains are only of the order of one part in 10 million, they amount to about a 30 cm displacement at the earth's surface. Since the earth rotates on its own axis, the moon revolves around the earth, and the earth revolves around the sun, the magnitude of the gravitational pull is constantly changing, causing a continuous deformation of the earth, dissipating a substantial amount of energy in the form of the heat. The source of this heat energy is the kinetic energy of rotation of the earth which it loses as it gradually slows down at the observed rate of 3×10^{-20} degrees/s/s, corresponding to a 3×10^{19} ergs/s rate of dissipation of energy.

The amount of solar radiation, 1.05×10^6 cal/cm^2/yr, is relatively insignificant compared to other energy sources.

The principal source of the earth's heat energy has been established as radioactive decay. All rocks contain small amounts of radioactive elements but the main heat producers are uranium, thorium, potassium, and their products. Heat production characteristics of these radioactive isotopes are listed in Table 2-6. Even the small amounts of them contained in the minerals of common rocks

(Table 2-7) appear to account for the bulk of the observed heat flow. A 14 km layer of granite alone would produce 1 μcal/cm^2 of flow. The more basic rocks, basalt, peridotite, olivine, and all similar ones buried below the granite are not very radioactive and do not contribute much to the overall heat flow. The greatest uncertainty as far as heat production is concerned is the distribution of radioactivity in depth. A reasonable assumption is that the largest share of the total heat is generated in the crust. Table 2-8 lists reasonable estimates of the contributions made by different layers of the earth to the total heat output.

Table 2-6. Heat from Radioactive Decay (Adapted from Bullard, 1973)

Isotope	Heat Generated (cal/g/yr)	Half Life (10^9 yr)	Reaction
U^{238}	0.70	4.50	$U \rightarrow Pb^{206}$
Th^{232}	0.20	13.9	
K^{40}	0.21	1.31	$K \rightarrow$ or $\begin{array}{c} Ca^{40} \\ A^{40} \end{array}$
$K_{ordinary}$	27×10^{-6}	—	
U^{233}	4.3	—	
$U_{ordinary}$	0.73	—	
U^{235}	0.03	0.71	

Table 2-7. Heat Production in Common Rocks (Adapted from Bullard, 1973)

Rock Type	Concentration of Elements			Heat Production (μcal/g/yr)			
	U ppm	Th ppm	K %	U	Th	K	Total
Granite	4.7	20	3.4	3.4	4.0	0.9	8.3
Basalt	0.6	2.7	0.8	0.44	0.54	0.23	1.21
Peridotite	0.016	0.004	0.0012	0.0012	0.001	0.0003	0.013

Table 2-8. Heat Production in Different Layers of the Earth (From Stacey, 1969)

	Total Heat Production for Complete Spherical Shell (ergs/s $\times 10^{19}$)	Radioactive Production per Unit Volume	Mean Density (g/cm^3)	Heat Production per Unit Mass (ergs/g/yr)
Continental				
Crust	20	200	2.7	104
Upper Mantle	8	8	3.35	3.34
Lower Mantle	3	1	5.15	0.27
Oceanic				
Upper Mantle	28	28	3.35	11.6
Lower Mantle	3	1	5.15	0.27

2.3 Transport and Distribution of Heat

Heat transfer is perhaps the most important of all geologic processes since it provides the bulk of the energy required to drive most of the others. The heat when converted into mechanical work provides that needed to move continents, to make mountains, to cause volcanoes to erupt, to supply water and steam to geothermal areas, and to form concentrated deposits of useful and valuable ores.

The three primary mechanisms by which heat is transferred are radiation, conduction, and convection. Radiation is effective only at high temperatures. Since radiative transfer is proportional to the fourth power of the absolute temperature, it is of little consequence to the earth. Conduction is a very slow process but it is the dominant one as far as the earth at large is concerned. Convection, however, also plays an important role on both an extremely large and smaller scale. On the large scale, the viscous mantle forms large convection cells on top of which ride the continental plates. On a smaller scale, the numerous fractures and fissures and porous alluvial deposits which lie near the earth's surface provide passageways for rapid circulation of the magmas and other fluids, especially the water and steam that furnish the volcanoes and geothermal areas with their heat and fluids.

In principal, convection is easy to understand. When a fluid is heated, it becomes less dense and tends to rise because of the bouyant forces acting in accordance with Archimedes principal, which specifies that a body will be buoyed up by a force equal to the weight of the fluid it displaces. Cool fluid, being more dense, immediately sinks to replace the warm fluid that has risen. If the fluid is free to circulate without restriction, as in a large reservoir, a definite pattern of circulation establishes itself. Often several well-defined cells form with fluid asending in some regions and descending in others. The earth's mantle is believed to be broken up into several such cells, each having dimensions of the order of several hundreds of kilometers. These cells are most likely the agents determining the observed pattern of surface deformation, particularly mountain building, continental drift, and sea floor spreading.

Very often, however, free circulation cannot take place because of constraints placed on the fluid. This is usually the case in volcanic and geothermal areas. Flow of the lavas, the volcanic gases, the water and the steam in the upper 5 to 10 km of the earth is restricted to specific channels, generally the abundant fissures, joints, fractures, faults, and beds of porous rock. In addition, such constraints can easily impede the attainment of thermodynamic and hydrodynamic equilibrium and cause highly unstable, sometimes catastrophic situations to develop—geyser and volcanic eruptions.

Both steady state and intermittent convection are the principal means by which energy is finally transferred to the earth's surface in geyser areas. But actually, most of the heat energy is transferred by conduction upward toward the earth's surface from its radioactive sources to the crust. The rate of heat transfer, q, which is very slow, varies from region to region depending upon the thermal

conductivity, K, of the ambient rock. The thermal conductivity is the constant of proportionality between heat flow and temperature gradient, dT/dz, thus:

$$q = K(dT/dz). \qquad (2-1)$$

It is apparent from Eq. (2-1) that high thermal conductivity indicates high heat flow for a given temperature gradient. Or contrariwise, as is the situation close to the surface of the earth where the outward heat flow constant varies from rock mass to rock mass, the temperature gradient must also vary to maintain constant heat flow. Heat flow from the surface of the earth is determined experimentally by calculating (dT/dz) from temperatures measured at two points about 2 m apart in a vertical hole and then combining the value of the field-measured temperature gradient with a laboratory measurement of the thermal conductivity of a core taken from the hole. Values of thermal conductivity vary by as much as a factor of about 6 among rocks (Table 2-9). Using the average observed heat flow value of 1.5 μcal/cm^2/s and an average thermal conductivity of 4×10^{-3} cal/cm·s when substituted in Eq. (2-1) gives an average temperature gradient of 25°C/km near the surface of the earth. Extrapolating downward, the boiling point of water is reached at about 3 km depth. Such a gradient is commonly observed in deep mines and deep drill holes. The gradient will be relatively high in rocks having low thermal conductivity and vice versa.

Heat flow is very uneven over the surface of the earth and many hot spots other than the volcanic and hot spring areas exist where temperatures as high as 175°C/km of depth may exist. Generally, they break out where the earth's crust has been severely disrupted as a result of tectonic movement.

The Yellowstone hot springs receive their heat from cooling magma at an approximate temperature of 600°C. Meteoric water which has circulated to great depth, as much as from 1500 to 3000 m, carries the heat to the surface which ultimately causes the geysers to play. It is most unlikely that this water

Table 2-9. Thermal Conductivity of Rocks

Rocks	K (cal/cm·s $\times$ 10^{-3})
Chalk	2.2
Schist	2 to 4
Limestone	5 to 7
Igneous rocks	6
Marble	7.5
Granite	8
Rhyolite	8
Dolomite	10
Dunite	12
Rock salt	13
Quartzite	16

ever attains the 600°C of the magma. Rather, heat is first conducted through a layer of rock overlying the magma with the water then taking up the conducted heat. If heated above 330°C, the water would tend to deposit dissolved silica due to an abrupt decrease in its solubility at temperatures above that. As a result, it might be expected to seal off its own convection system.

An alternative view, supported to some extent by drilling at Yellowstone, is that even if the depth to which ground water penetrates is indeed sharply limited by temperature, there then lies below this depth and extending down to the magma a zone of rock permeated with pores, cracks, and possibly fissures, filled with superheated steam that supplies heat directly to the geysers.

In the fresher volcanic areas of Iceland, New Zealand, and Kamchatka, convection of volcanic gases given off by the cooling liquid magma upward through fractures and joints seem also to contribute heavily to the heating of the geyser waters. If the water-carrying passages are not too obstructed, large bubbles of steam may surge upward only to condense quickly when they reach cooler water, to which they transfer their heat.

In Iceland and New Zealand, the distribution of hot springs is closely related to the pattern of visible faults. Groups of craters, fumaroles, and hot springs form along approximate straight lines. But no such close, striking relationships between faults and hot springs are seen at Yellowstone. At Norris and Upper Geyser Basin, penetration, circulation, and egress of water and steam seem to take place along surficial local joints and shrinkage cracks, not proper faults. Drilling there strongly indicates that the hot water channels as a rule go vertically down, or almost so, although perhaps a bit crooked.

2.4 Storage of Heat

Cooling magma bodies supply the relatively small volume of the earth's crust occupied by hot springs with the large amount of heat they discharge. The heat of fusion of lava lies between 75 and 100 cal/g; the specific heat of lava is about 0.2 cal/g. The temperatures of lava normally range from 850°C to 1200°C so that on cooling and crystallizing from a temperature of 1000°C, it would liberate about 400 cal/g of heat. Solidification of magma at the rate of 1 km^3 every 10 years would adequately supply Yellowstone's heat requirement of 1200×10^6 cal/s.

All magma bodies do not erupt as volcanoes. Some form plutons, plumes of upwelling magma that never reach the surface. Magma is less dense than igneous rock so that there is a general tendency for a magma body, once formed, to be buoyed upward. Some simply encounter a higher resistance near the surface than they can overcome and never see the light of day. Such bodies, mostly now cool, are numerous; 100 to 200 lie beneath a 100×600 km area of the Sierra Nevada in California. Yellowstone derives its heat from a presumably still molten, approximately 55×85 km, pluton buried at a depth of about 5 km.

A very complicated series of events begins to take place as magma moves upward toward the surface of the earth. Gases and liquids held in solution by

virtue of the high pressures will be released as pressure falls and the magma will begin to crystallize as its temperature falls. Lava is a complex mixture of oxides and silicates, with the potential of forming many different solid substances depending upon the pressure, temperature, water, and volatile content. Elements and combinations of elements that do not form high melting point compounds are usually left over as gases and liquids as the magma crystallizes. Hot water and steam, which constitute about 5 percent by weight of most magmas, will be the dominant expelled fluids. By steam distillation they will carry away many other materials, even some of those present in only small quantities such as mercury and silver, which may be deposited elsewhere in mineable quantities. The elemental sulfur deposits found around the geysers at Whakarewarewa, the sulfur dioxide gas, and some of the steam in many hot spring waters were generated in this way.

The quantity of heat carried away by the water and steam depends upon he temperature and pressure within the magma. Increase of temperature increases heat content. Increase of pressure decreases it. For magmatic steam evolved at constant temperature, the deeper the magma the less the heat content of the steam. Assuming that the heat content is maximum, the total heat given up by a km^3 of magma cooling from 1100°C to 600°C would be

	Cal $\times$ 10^{15}
Latent heat of fusion of magma liberated	2.0
Heat loss from solid magma cooling to 600°C	3.4
Heat carried away by 5 percent of water	1.1
Total	6.5

The passage of magmatic steam upward from depth provides a very attractive mechanism for supplying heat to geyser systems.

In geyser areas, the rock mass lying close to the pluton or hot lava flow is generally completely saturated with water, the amount held by the rock depending upon its porosity. Some rocks, for example most sandstones, are extremely porous, containing as much as 20 percent void space, whereas others like dolomite are 97 percent solid mineral matter.

Relatively shallow, down to tens of meters at most, thermal and hydrological conditions dominate geyser phenomena. Shallow subsurface temperatures have been measured at most of the geyser areas. Rather extensive research drilling programs have been undertaken at Yellowstone with 13 research holes being drilled to depths ranging from 65 m to 330 m.

The Yellowstone drillings identified a near surface zone, ranging from less than 30 m thick to more than 75 m characterized by a large approximately linear thermal gradient. The zone of originally mainly highly permeable rhyolitic gravel has now lost most of its permeability through hydrothermal alteration and deposition of silica and silicate minerals. The first aquifer, water-saturated mass of rock in which significant convective flow occurs, lies just below this zone in the

extensively fractured rhyolite first identified by C. N. Fenner in his 1926 drill
hole at Upper Geyser Basin.

Some of the drilling data are summarized in Table 2-10. Variation of temperature with depth followed the reference boiling point curve (Fig. 3-6a). This is a
curve calculated on the assumption that the water at any position in a vertical tube
will boil as soon as its vapor pressure, which increases with temperature, reaches
the hydrostatic pressure exerted by the overlying water. The highest temperature,
238°C, was recorded at the bottom of the deepest hole, the 330 m deep Y-12 hole
in Norris Basin. The shallower, 60 to 150 m deep holes in Upper Basin were only
about 170°C whereas the 160 m deep holes in Midway and Lower Basins reached
a little over 200°C.

One surprising result, for which the data are not shown in the table, was that
the vertical water pressure gradients in all holes except one exceeded by 110
percent that which would be exerted by a simple column of water. It was as high
as 147 percent in one hole in Lower Basin. The temperatures, however, were
never in excess of that permitted by the ambient pressure. These very high
temperatures and pressures alone would readily account for the abundance of
geysers in Yellowstone.

Drilling is the more active parts of the Steamboat Springs area showed that
there exists an upper zone in which the temperature increases rapidly with depth,

Table 2-10. Temperature Data Obtained from Drill Holes in Yellowstone Geyser Basins
(Adapted from White et al., 1975)

	Total Depth Drilled (m)	Depth to Bedrock (m)	Maximum Measured Temperature (°C @ m)
Upper Basin			
Y-1	66	64	171 @ 66
Y-7	74	53	143 @ 72
Y-8	153	55	170 @ 52
Fenner Hole	124	67	180 @ 55
Midway and Lower Basins			
Y-2	157	—	204 @ 157
Y-3	157	—	193 @ 134
Y-4	210	—	195 @ 209
Y-13	142	—	203 @ 142
Y-5	164	—	170 @ 163
Norris Basin			
Y-9	248	—	196 @ 248
Y-12	332	—	238 @ 332
Fenner Hole	75	—	205 @ 75

an intermediate zone ranging from 30 to 90 m in depth where temperatures are very close to but a few degrees below the boiling point curve for pure water, and a deep zone where the temperature levels off at about 170°C, with no further increase at greater depth.

Table 2-11 is a tabulation of the approximate size, maximum recorded subsurface temperature, and total heat flow in geyser areas in various parts of the world. In general, the thermal data indicate that deep subsurface temperatures of at least 150°C to 160°C are necessary to support natural geysers near the surface.

Basing conclusions on geochemical analysis and heat content, studies of the geyser and hot spring waters, and on water from numerous drilled holes, it is believed the depth to Yellowstone's deep geothermal reservoir is more than 2 or 3 km, but no more than 5 km, the established location of the pluton, and that the temperature of the underground water in the deepest reservoir ranges from 300°C to 350°C. This estimate assumes that vapor saturated water overlies the reservoir. Geologic indications are that the rock mass of the reservoir is comprised of ash flow tuffs, rhyolite flows, and sediments capable of providing abundant storage of hot fluids. The aquifer underlying the entire Yellowstone caldera is covered in some regions by impermeable lake sediments that restrict upward and downward circulation to the surface.

The specific heat of the rock mass and its capacity to store heat depends upon its composition, porosity, and degree of saturation, which vary widely. It turns out, however, that most of the in situ masses located in geothermal areas have

Table 2-11. Extent and Thermal Properties of Geyser Areas (Adapted from White, 1965)

Area	Approximate Size (km^2)	Maximum Recorded Temperature (°C)	Total Heat Flow (10^6 cal/s)
Japan			
Atami	5	180	16
Onikobe	~80	185	?
New Zealand			
Wairakei	7	266	130
Orakei-Korako	~5	boiling*	130
Rotorua	?	>160	?
United States			
Steamboat Springs	5	187	7
Beowawe	~3	207	?
Yellowstone (total)	70	205	500
Norris Geyser Basin	~3	238	8
Upper Geyser Basin	~10	180	90

* Temperature measured at surface

very similar volumetric specific heats, increasing nearly linearly with temperature:

Volumetric Specific Heats (cal/cm^3/°C)	Temperatures (°C)
0.6	100
0.7	300
0.8	800

The geyser basins appear to be small zones of intense upflow of very hot water spotted here and there in much larger areas of diffuse downflow. Noneruptive hot springs dissipate the energy delivered to them by continuous means such as surface boiling, continuous streaming of steam bubbles, evaporation, and conduction of heat into cooler surrounding rock. Geysering is a nonsteady state phenomenon where the energy is not continuously dissipated rapidly, leading to the temperature buildup that precipitates the catastrophic events that eventually rid the geyser of its excess energy.

2.5 Heat Efflux

Heat can be discharged from a thermal area in the following ways:

- flow through the soil;
- loss from water surfaces;
- escape of steam from fumaroles;
- overflow and geysering of hot water; and
- underground seepage of hot water into nearby lakes and rivers.

As an example, Table 2-12 lists the partitioning of heat flow from the Wairakei thermal area when it was still dotted with geysers, hot springs, and fumaroles, before it was extensively developed for geothermal power. The temperature gradient at Wairakei ranges from 3°C/m to 15°C/m.

The heat brought to the surface by the hot water discharged from the hundreds of geysers and hot springs in Yellowstone has been calculated and found to be about 220,000 kg cal/s, enough heat to melt three metric tons of ice per second. The heat coming out of the 47 km^2 area of Upper, Midway, and Lower Geyser Basins alone is enough to melt a 4.5 m thick layer of ice per year; the amount of heat given off by the Upper Geyser Basin is 800 times the amount given off by nonthermal areas of the same size. Records for the past 100 years indicate that the underground temperatures have not cooled measurably in that time. Indeed, geologic studies there indicate that very high heat flows have continued for at least the last 40,000 years.

The above estimates are based mostly on detailed measurements of aggregate

Table 2-12. Heat Flow from the Wairakei Area (Adapted from Healy, 1953)

	Steam (cal/s)	Water (cal/s)	Combined (cal/s)
Hot springs and geysers	11,500	41,500	53,000
Karapiti fumarole	3,000	—	3,000
Steaming ground	51,000	—	51,000
Conduction	—	5,000	5,000
Seepage to Waikato River	—	18,000	18,000
Total	65,500	64,500	130,000

discharge of water, hundreds of temperatures measurements made in the springs, and to a lesser extent on chemical studies involving measurement of the chloride content of the waters flowing out of Yellowstone. Table 2-13 lists the heat carried away by each of the important groups of springs. Each entry was obtained by multiplying the volume of water discharged by the group by the difference between the average temperature of the waters flowing from the group and the average air temperature, which is 5°C there. The total measured rate of discharge of water, 3160 liters/s has been increased to 3350 liters/s to take into account those springs not listed. The average maximum temperature of the hot water discharged is 76°C.

No reliable estimate has been made of the total efflux of heat at Yellowstone including fumarole, steaming ground, conduction, and underground seepage. Using the Wairakei estimate of total heat flow (Table 2-12) as a guide, the additional heat might possibly amount to even more than that discharged as hot water.

It is a fairly straightforward procedure to calculate reasonably accurately the rate of heat discharge in the form of steam and hot water from a specific geyser. This has been done at three Yellowstone geysers: Old Faithful, Narcissus, and Solitary. The primary field data needed are the distribution of temperature within the geyser at the time of an eruption, the behavior of the geyser during it, and how often eruptions occur.

The seasonal average interval between eruptions of Old Faithful is above 66.5 min. It has been found to erupt when its water temperature reaches 112°C; the ambient boiling point at the geyser opening, 2204 m altitude, is 93°C. When 4 percent of the water has flashed into steam, the energy absorbed in the conversion process will cause the temperature of the remaining water to have fallen to 93°C and the boiling stops. Temperature measurements show that the temperature of the water decreases uniformly and essentially linearly from 112°C at the beginning of the eruption to 93°C at the end (Fig. 2-6). While the total quantity of water ejected during an eruption has never been carefully monitored, it is believed to average about 50,000 liters. At least 95 percent of geyser water being meteoric in origin enters the geyser's catch basin at mean ambient air temperature. The total heat required to heat the 50,000 liters of water from 5°C to 112°C

Table 2-13. Heat Carried Away by Hot Spring Waters of Yellowstone (Adapted from Allen and Day, 1935)

Springs	Discharge (liters)	Average Temperature of Water (° C)	Heat (kg cal/s)
Upper Basin	518	90	44020
Excelsior Geyser, etc.	223	88	18510
Prismatic Lake	35	63	2030
Rabbit Creek 1	57	85 (?)	4560
Rabbit Creek 2	52	50 (?)	2340
White Creek and Firehole Lake	342	85 (?)	27360
Fountain Group	40	89	3360
River Group	66	81	5016
Sentinel Group	18	80	1350
Fairy Creek (portion)	54	85	4320
Imperial Geyser	43	85	3440
Nez Perce headwaters	37	51	1702
The Thumb	47	81	3572
Violet Creek	46	80 (?)	3450
Hot River	640	53	30720
Mammoth	55	62	3135
Clearwater	14	89	1176
Bijah	4	84	316
Norris Basin	39	85	3120
Chocolate Pots	4	55	200
Artist's Paint Pots	9	60 (?)	495
Geyser Creek	9	86	729
Terrace	93	62	5301
Shoshone Basin	66	89	5544
Lewis Lake	28	75 (?)	1960
Heart Lake	104	88	8632
Snake River	13	60	710
Polecat Creek	11	59	594
Alum Creek	245	85 (?)	19600
Total	2912		207262

is 5.35×10^9 cal. Since this heating operation takes place on the average once every 66.5 min, the rate of heat input from the earth to the geyser water is 1.34×10^6 cal/s. Essentially no water returns to the geyser so that this figure is also the rate of heat flow from Old Faithful.

Narcissus, a quite different type of geyser, is a fountain geyser having an

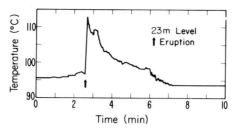

Figure 2-6. Temperature of ejected water during eruption of Old Faithful.

hourglass-shaped vertical cross section. It is about 3 m in diameter at the top, narrows to a 0.6 m diameter pipe about 0.2 m down, and then opens up again toward the bottom. The total depth is 6.5 m. It fills slowly between eruptions which occur at alternating intervals of 2 and 4 hr. The eruption consists of a succession, sometimes more than 300, of detonating explosions, a few throwing water and steam to a height of 10 m. An eruption takes place when relatively light superheated water at the bottom of the hourglass becomes unstable and rises to the top where a fraction of the overturned bottom water flashes at the top into steam. Plots of the measured temperatures as a function of time in both the lower and upper chambers are plotted in Fig. 2-7. Such an eruption ordinarily takes place when the temperature of the bottom water heats up to 105°C, 12°C above the ambient boiling point at the surface The temperature is nearly constant in the bottom chamber for a height of about 2 m.

Flashing of the superheated water into steam is the principal mode of effusion of heat from Narcissus. The water run off is essentially negligible. The volume of the top reservoir is about three times that of the bottom. The temperature of the top water just before mixing occurs is about 90°C when the elapsed time since the

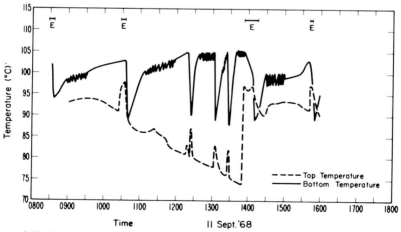

Figure 2-7. Temperatures in reservoirs at Narcissus as a function of time. Upper curve for bottom reservoir; lower curve for top reservoir. E indicates eruption.

last eruption is only two hours, and 75°C when it is four hours. As the bottom water mixes with the top water, the explosions and mixing cool the bottom water to the ambient boiling point of water, 93°C, and heat up the top water to the same temperature. The volume of the lower reservoir is 3.0×10^6 ml, thus at a temperature of 105°C, its heat content in excess of the boiling temperature, 93°C, is 36.0×10^6 cal. Part of this heat is used to heat the water in the upper reservoir and eventually is taken up by the atmosphere as the top cools down between eruptions. The remainder goes into the formation of the eruption's numerous steam explosions. Averaging the eruption intervals at 3 hr, the total rate of heat flow is 3.3×10^3 cal/s.

An eruption lasts from 5 to 15 min, averaging 10 min, during which an average 200 explosive bursts occur, the number of bursts being almost directly proportional to the interval between eruptions. Each explosive bubble carries away about 1.8×10^5 cal of heat.

Solitary Geyser, Upper Geyser Basin, is also a fountain geyser. It has a 20 m diameter pool, constantly filled with water above a deep, narrow vertical tube. The eruption consists of a succession of detonating steam bubbles (Fig. 2-8). Each explosion is the result of a mass of water rising from below and suddenly flashing into steam. Extensive temperature measurements made down to a depth of 270 m (Fig. 2-9) definitely establish that the water is heated from below by a series of injections of 160°C to 170°C water occurring at intervals of about 1 min. For a few minutes after each eruption, fairly extensive circulation takes place down to about the 140 m level. This is followed by a quiescent period during which the water stagnates and at each level stays at about a constant temperature. Suddenly a blob of light superheated water rises to the top where it turns to steam, and cooler top water drops downward to replace it. After several such overturnings, the eruption episode stops, the water in the basin becomes stable, and a new heating and circulation cycle begins.

Heat flow from Solitary is difficult to estimate since there is no practical way to determine the volume of water involved. A very rough estimate can be based on the size and number of steam bubbles. The average interval between eruptions is about 7 min, each eruption lasting about 1 min during which something like 10 steam explosions occur. From calculations on Narcissus, each blob of its superheated water which flashes into steam contains about 1.8×10^5 cal of excess heat. The temperature of the water in Solitary's basin just before an eruption is 110°C, 7°C or 42 percent hotter with respect to the 93°C boiling point than that in Narcissus. Thus a reasonable extrapolation is that each explosion in Solitary represents 2.6×10^5 cal of energy. These rather crude assumptions indicate that the heat flow from Solitary is of the order of 6.2×10^3 cal/s, about twice as much as from Narcissus.

The magnitude of these heat flows are, of course, many times larger than can be accounted for on the basis of the 1.5μcal/cm^2 world average, but they are also much greater than the high values of 40μcal/cm^2/s average which geothermal areas such as Yellowstone would sustain. Assuming that 40μcal/cm^2 obtains in the Yellowstone geyser basins, the heat flowing from Old Faithful must have

Figure 2-8. Solitary Geyser during eruption.

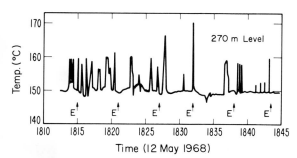

Figure 2-9. Temperature as a function of time at a depth of 270 m in Solitary. E indicates eruption.

been extracted from an area of 3.4 km²; for Narcissus, the area would be 8.3 × 10^{-3} km²; and for Solitary 16.0 × 10^{-3} km². These values are compatible with visual observations of the region. Old Faithful lies at the head of a large catch basin whose dimensions could easily encompass an area about 1 km in diameter. It shares this basin with several other geysers and hot springs. Narcissus and Solitary are both isolated geysers whose performance might be expected to be affected by only very local anomalies.

Heat flow from these three geysers constitute only a small part of the total, 840 × 10^6 cal/s, heat flow from all of the Yellowstone region. However the calculations certainly point to the tremendous concentration of energy in a particular geyser. For example, the power, energy per unit time, of Old Faithful alone, 1.34 × 10^6 cal/s or 5000 kw, is enough, if converted with 100 percent efficiency to electrical power, to provide the normal electrical needs for a town of 5000 persons.

CHAPTER 3

Fundamentals of Geyser Operation

3.1 Essential Elements of a Geyser

Geysering is a captivating and mysterious phenomenon. Almost anyone who watches a geyser erupt speculates on its mode of action. The essential elements of a geyser are a reservoir and associated plumbing system in which water can be stored and heated, an adequate supply of water, and a source of heat. While other factors such as gases, earth stresses, dissolved chemicals, and climate can and often do influence geyser action, it is not essential that they participate.

Knowledge of the precise natures of geyser reservoirs is very limited simply because they reside mostly out of view underground. Probings to any appreciable depth are difficult because of internal obstructions and there has been no drilling close to the large geysers. Only in one case, the extinct Te Waro Geyser of Whakarewarewa, has man been able to climb down inside the reservoir itself. Although the opening to Te Waro's original reservoir is now covered with sinter deposits, in 1921 the external vent was "a circular hole with . . . typical round sinter edges of a geyser, just large enough for a man to squeeze his body through". Mr. Martin's investigation has been described:

> The bottom of the shaft is 15 ft [4.6 m] below the surface, and it opens out into a chamber 12 ft [3.6 m] long and 9 ft [2.7 m] high, as shown in the following diagram (Fig. 3-1). In the floor of the chamber are two fissures, one of which, F, is supposed to connect with Pohutu Geyser and whence come the rumblings of fiercely boiling water. Mr. Martin suggested that this curious cavern-formation was due to the action of hot siliceous water and steam on an ordinary fissure passage, the water depositing silica on the walls and floors, and steam eroding the vaulted roof and forming a cavern.

Usually a geyser reservoir is a compact, well defined cavity or an interconnected array of cavities, the walls of which are, for the most part, lined with an

50

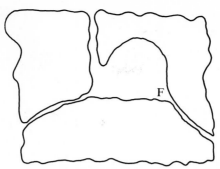

Figure 3-1. Cross section of Te Waro Geyser. (Adapted from Lloyd, 1975.)

impervious layer of siliceous sinter or geyserite that has been deposited by the geyser waters. It is now generally believed that most large voids and geyser tubes are effects rather than the original causes of geyser action. During an eruption, the waters in the reservoir may be merely agitated, partially erupted, or totally emptied, making it difficult from surface observations to estimate its capacity.

One can conjure up a reservoir system to fit almost any type of eruption and many investigators have done so during the past 150 years of geyser studies. Geyser theories have been proposed and numerous models constructed to simulate the actions of natural geysers and to prove or disprove certain theories. The most recent studies were reported in 1978. All of the models, which differ greatly in design, function. Perhaps those that did not were never reported!

Neglecting those instances where geysers are interconnected, six generic types of reservoirs can account reasonably well for most geyser action. These are shown in cross section in Fig. 3-2.

Riverside, Old Faithful, and Beehive, which play sustained jets to considerable heights at more or less regular intervals out of raised cones, probably have single standpipes a few hundred meters long (Type A, Fig. 3-2). At the opposite end of size range, Vixen, a small geyser in Norris Basin, erupts every minute from a 10 cm diameter geyserite pipe only 3 m long.

Deep, narrow shafts (Type B) spawn explosively violent, short-lived eruptions. They start abruptly, develop rapidly, and empty the pool quickly. Round Geyser in Yellowstone puts on a good show, playing to an average height of 25 m for a minute or so every 8 hr.

Some geysers, such as Strokkur, do not build cones but their standpipes have slightly raised rims around the openings and are surrounded by pools of water (Type C). Strokkur's tube is 13.5 m deep and funnel-shaped, 8.3 m in diameter at the top and only 26 cm in diameter 8.3 m down. Its action is quite violent and intermediate between that of a purely columnar geyser like Old Faithful and a pool geyser like Narcissus. Every minute or so a large bubble, a meter or two in diameter, rises to the surface and breaks explosively with the evolution of steam. Some explosions are much larger than others. Although the basin remains nearly full most of the time, the water level lowers several centimeters after each

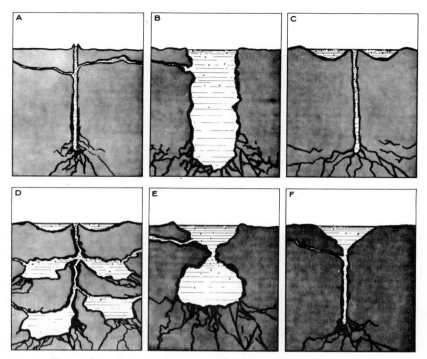

Figure 3-2. Cross sections of generic types of geyser reservoirs.

explosion. Every 10 to 15 min the geyser erupts, throwing steam and a small quantity of water to a height of 20 m or more (Fig. 3-3). The eruption is short-lived, a matter of only a few seconds.

Two large fountain geysers, Grand and Great Fountain, erupt as a series of powerful and sustained fanlike jets of steam and water, each burst being separated by a short period of inactivity. There are fewer pauses in the action of Great Fountain but they are longer in duration, sometimes as much as 30 min. A very logical explanation for this type of activity is that each geyser has several interconnected underground reservoirs, one emptying after another, somewhat as shown in the figure as Type D. Other configurations of pool or fountain geysers look something like Type E for Narcissus, and F for the Great Geysir.

Temperature soundings, recorded seismic signatures, and visual aspects of play are all additional clues as to the nature of the reservoir and its associated water and heat sources. These suggest that very often there is a complex network of two systems of channels, one shallow system feeding the bulk of the water to the reservoir and another deeper system feeding a much smaller quantity of extremely hot water and perhaps also hot volcanic gases.

Most of the water in a reservoir is trapped meteoric water, rain or melted snow, which has followed one of several paths to it. Geysers are located frequently along banks of rivers from which certainly some of their water is derived

Figure 3-3. Strokkur in full eruption.

(Fig. 3-4). Part of the water has made a long circuitous journey downward through heavily fractured and altered hot rock where it is first heated and then convected back to the surface. It is possible that some of the water that enters the reservoir is magmatic.

The bulk of the heat supplied to the geyser is by injection of hot water or steam directly into the reservoir and channels already filled with hot water. Heat conduction through the walls of the reservoir plays only a minor role.

The most definitive indicators that such injections are occurring are deep-down temperature measurements. At the 270 m level in Solitary, these injections are intermittent and sporadic but have a high frequency of occurrence compared to the number of eruptions and hence only produce an average effect (Fig. 2-9). The temperature of the injected water, about 170°C, is not especially high, certainly well below the boiling point of about 230°C at the pressure produced by a hydrostatic head of 270 m. Water of this temperature must rise toward the surface where it will mix in the other water before it can be effective in precipitating an eruption.

In Old Faithful, evidence indicates that mixing of water occurs below the 175 m level, ruling out the entrance of surface waters at very shallow levels. The bulk of the hot water is apparently injected as large surges spaced 20 to 30 min apart. The first of these after an eruption provides the energy for a series of steam

Figure 3-4. Riverside Geyser at beginning of steam phase.

bubbles and causes a series of seismic signals to begin. Succeeding surges supply additional energy. Any one of the surges could possibly trigger an eruption although a definitive causal relationship has not been established.

From temperature-depth measurements made on 25 geysers and springs in Yellowstone, it was found that in 16 of them a maximum temperature was reached before hitting bottom and the temperature increased very little at lower depths, suggesting that in many cases effusion of hot water takes place somewhere higher up from the bottom.

In order to appreciate the behavior of geysers, it is necessary to understand:

- why a geyser erupts;
- how the energy is supplied;
- how an eruption is initiated;
- how an eruption is terminated;

- why the behavior of geysers differ so much from each other; and
- why a single geyser may come into existence, change its behavior pattern, or cease.

Mackenzie after his original visit to Iceland published the first scientific theory of geyser action. He assumed that an eruption was caused by the expansive force of steam that accumulated in an invisible underground cavity. This idea was popular with some of the early investigators. Krug von Nidda visualized several branching cavities where steam collected, eventually forcing the water out and up through the geyser tube, thereby emptying it and opening a passage for the steam. Allen and Day felt that in the sinter-lined, steam-tight cavities, expansion and contraction of the steam contained in them caused the fluctuations in the surface level of water observed in many geysers. While none of these theories is universally applicable, each is partially correct.

Bunsen developed a theory intended to explain the action of the Great Geysir only. Although accepted for a long time, it does not seem to apply to the action of the Great Geysir but is fairly realistic in describing the action of a simple columnar geyser. Bunsen believed, on the basis of his temperature measurements, that the temperature at all points in the Great Geysir rose steadily from one eruption to the next with the temperature approaching the closest to the boiling point curve to be at the midpoint in depth, about 20 m. Consequently, boiling, which could be induced by the decrease in hydrostatic pressure caused by a 1 or 2 m overflow of water, would begin there. The steam generated by the boiling would further reduce the pressure; boiling would then begin to move downward, generating more and more steam and causing a full-fledged eruption.

Much has been learned in recent years concerning the principles of geyser action. From these, it is now possible to derive fairly realistic analytical solutions for typical geyser systems and to predict probable behavior patterns on the basis of measured physical parameters of the geysers.

3.2 Properties of Water and Steam

No matter which theory is proposed, it is obvious that two of the basic determinants in geyser behavior are the hydrologic and thermal properties of water and steam. Water transports and stores temporarily the heat energy that powers a geyser. In addition, it possesses several important properties upon which geyser action critically depends. It is a fluid, and, similar to most fluids, its specific gravity decreases as its temperature increases. Since a geyser's heat sources are generally nonuniformly distributed, high temperature gradients and hence corresponding high density gradients develop within the system. These density gradients are the driving force that establishes and maintains circulation throughout the whole geyser basin, both the large saturated rock masses and the individual geyser reservoir systems. The boiling point of water is pressure sensitive, increasing with increasing pressure. In geysers, boiling is temporarily suppressed

by the pressure of overlying waters, the hydrostatic head, and excess heat is stored only to be catastrophically released at a later time. And finally, the approximately 1500 fold increase in volume which occurs when liquid water varporizes into steam is capable of performing a prodigious amount of mechanical work, specifically, the propulsion of a large amount of hot water and steam at high velocity from the orifice of a geyser.

Deeply circulating thermal systems are large-scale convective fluid systems. Cold, relatively high-density water that falls on the surface percolates downward to replace heated, less dense water that is thus buoyed upward. The deep circulation patterns with dimensions of the order of a few thousand meters, take place principally through faults and fractures. Close to the surface, especially in the upper 60 to 70 m, many interconnected channels develop which often enlarge upward, and the porosity of sediments and fault breccia will increase. At shallow depths, the density of water varies markedly with depth due to high temperature gradients existing there. Consequently many secondary convection patterns develop. The velocities of flow of the fluids have been calculated to be of the order of a few centimeters a day.

The free circulation occurring in many hot pools is generally not present in geysers. The violent eruptive behavior of geysers depends on the fact that constrictions in the reservoir plumbing restrict circulation (Type E, Fig. 3-2), allowing instabilities to develop where denser cooler waters overlie lighter warmer waters. Excessively high temperatures are required to create the higher than normal density differences needed to develop the buoyant forces adequate to break through the constriction and erase the unstable situation. Rapid massive circulation takes place during this time, often exposing superheated water to lower pressures where it is likely to burst explosively into steam.

The heat energy content of a substance or a system of substances is referred to as *enthalpy*, E. Most thermal processes involved either decreases or increases in the enthalpy of systems or transfer of enthalpy from one part of the system to another. When the enthalpy of a substance is increased by the addition of heat, its temperature may rise or its temperature may remain the same, the substance simply undergoing a phase transformation such as melting or boiling. Substances vary greatly, by more than a factor of five, in the amount of heat required to effect phase transformations and temperature changes. The amount of heat required to increase the temperature of a unit mass of a substance one degree is defined as its *specific heat*. Pure water has the highest specific heat, c, of any substance. A commonly used basic unit of heat energy is a *calorie*, cal, defined as the amount of heat required to raise the temperature of one gram of water 1°C. The *British Thermal Unit*, BTU, is the amount of heat required to raise the temperature of one pound of water 1°F, equivalent to about 1/4 cal. The specific heat of water varies slightly with temperature but not enough to affect appreciably any of the geyser calculations referred to here. Under standard sea level air pressure, 760 mm of mercury, water boils at 100°C (212°F). The conversion of liquid water to steam requires a relatively large amount of energy, 540 cal/g, defined as the *heat of vaporization*, σ. Boiling will stop unless heat is continually

added to the water to make up for the heat carried away by the steam, or unless the water is in a metastable superheated condition with its temperature higher than its normal boiling point. The steam bubbles when they first form are very small and uniformly distributed through the liquid. Their initial smallness results from the fact that a steam bubble can derive its heat only from its immediate surroundings. Superheated water will suddenly begin to boil violently, frequently internally and explosively, forming thousands of small bubbles of steam, using up all of the excess heat to do this, while the surrounding liquid cools down to its normal boiling temperature.

A very unusual phenomenon exhibited by many hot springs is the sustained presence of superheated water at the surface. The degree of superheat varies from a fraction of a degree to 3°C and, rarely, more. It always persists as a blob, recently migrated from depth, and never covers the surface uniformly. Usually the presence and extent of the blob can be observed visually but an almost sure test is to toss some foreign matter such as sand into the spring. The superheated water will explode into violent boiling, leaping up sometimes a few meters, hissing and roaring. Soon the temperature of the water drops to its normal boiling point. The earliest Icelandic travelers used to toss sod into the major geyser pools to make them boil up. The temperature of the water in springs that constantly churn is almost always above the ambient boiling point.

Water boils when its *vapor pressure,* p, pressure exerted by the water as it evaporates to form a bubble, which increases with increasing temperature, is just equal to the ambient pressure. Thus when the ambient pressure is lower, as at higher elevation, water at the ground surface will boil at a lower temperature. Deep within reservoirs, where the ambient pressure is higher, the water must attain a higher temperature before it will boil. Change in boiling point as a function of altitude is plotted in Fig. 3-5. Water boils at 100°C in the geyser areas of New Zealand, Iceland, Japan, and Kamchatka, which are all approximately at sea level. Yellowstone is at an elevation of 2200 m where water boils at a much lower temperature, 93°C. This lower boiling point adds to the ease with which the geysers can erupt since steam will form at a relatively much lower temperature.

The weight of superincumbent waters inhibits the development of most boiling hot springs and all geysers. The boiling point of pure water as a function of depth below the surface is plotted on a small scale to great depths in Fig. 3-6a. Figure 3-6b is a similar plot showing more detail at shallow depths. Down to about 10 m, the gradient is roughly 0.6°C/m. The curve breaks sharply at a depth of about 300 m and then straightens out at about 800 m with a much more gradual gradient, 0.02°C/m. Curves for mineralized waters do not differ appreciably.

The depth versus boiling point curve is not difficult to interpret and is extremely helpful in understanding geyser action. Assume that a small mass of water whose temperature is 110°C resides at a depth of 6 m. Since its temperature is lower than the ambient boiling point, the water will be in liquid form (Fig. 3-6 b); however it is superheated with respect to ground surface. Its temperature will

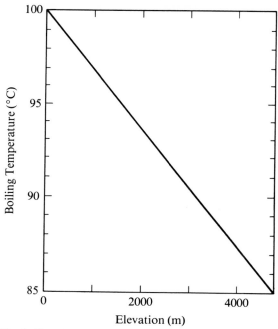

Figure 3-5. The boiling point of water as a function of elevation above sea level.

remain constant (the vertical line, ab) and it will be in liquid form until it reaches the depth corresponding to its crossing the boiling point curve. At that point, 3.7 m in depth, it vaporizes, gathering the heat to do so from its surroundings, and forms a bubble. As the bubble moves on upward, its temperature will decrease since it is in contact with cooler fluids. Either one of two things can happen. It will give up all of its excess heat to its surroundings and collapse, or the bubble will survive its journey through the cooler water, finally arriving at the surface where it bursts.

The amount of heat acquired from the earth by the water in one of Yellowstone's boiling springs by virtue of its residence in the earth can be calculated readily. The annual average mean air temperature there is 5°C. Thus, since precipitation is fairly uniform throughout the year, it can be assumed that the meteoric water which falls begins its journey through the earth at this temperature. At Yellowstone water boils at a temperature of 93°C. Its increase in enthalpy, ΔE, has therefore been its change in temperature times the specific heat of water or

$$\Delta E = (93 - 5) \times 1 = 88 \text{ cal.}$$

If the spring boils constantly, then it must be continually being supplied additional heat, most likely by small quantities of superheated water. Assuming that

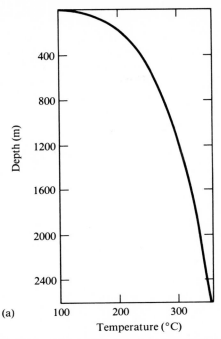

Figure 3-6a. The boiling point of water as a function of depth of overlying water.

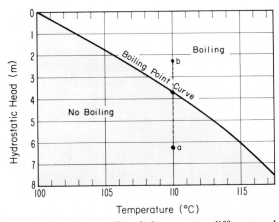

Figure 3-6b. Same as Fig. 3-6a except to different scale.

this latter water enters at a temperature of 110°C, the amount, m_w, of it needed to convert 1 g of the 93°C spring water into steam is given by

$$m_w (110 - 93) = 540$$

or

$$m_w = 32 \text{ g.}$$

The relative amount of steam formed when a given quantity of superheated water boils, releasing its excess heat, can be calculated in a reasonably straightforward manner. Let E_1 be the enthalpy of 1 g of the superheated water having a mean temperature of t_1, and E_2 the enthalpy of 1 g of the residual liquid which, as a result of the enthalpy, σ, carried away by the steam, has lowered its temperature to t_2, the ambient boiling point. Assuming that the evaporation and cooling process take place adiabatically, without gain from or loss of total energy, then

$$E_1 = E_2 (1 - x) + \sigma x \qquad (3\text{-}1)$$

where x is the fraction of water transformed to steam. Solving Eq. (3-1), x is given by

$$x = (E_1 - E_2)/(\sigma - E_2).$$

For 100°C water cooled to 100°C, $(E_1 - E_2)$ is approximately 10 cal/g. $\sigma - E_2$, is close to 540 cal/g, the heat of vaporization of water at 100°C under standard conditions. Thus,

$$x \approx 1.85\%.$$

The results of more precise calculations taking into account variations in specific heat with temperature for two values of t_2, 100°C and 110°C, are listed in Table 3-1 for superheated temperatures up as high as 310°C. At this temperature almost exactly one-third of the liquid flashes into steam. For a high altitude region like Yellowstone, a reasonably good estimate can be obtained by simply appropriately sliding the temperature scale, treating 110°C superheated water as if it were at 117°C relative to a 100°C boiling point.

Bubbles are created below the surface as steam forms, usually as a very large

Table 3-1. Properties of Water and Steam (Adapted from Golovina and Malov, 1960)

Temperature (°C)	Enthalpy of Water (cal/g deg)	Specific Volume of Steam (ml/g)	Enthalpy of Steam (cal/g deg)	wt % of Evaporated Liquid for the Evaporation Temperatures of 100°C	110°C
100	0.312	1674	1.756	—	—
110	0.339	1210	1.728	1.87	—
115	0.352	1050	1.714	2.77	—
120	0.365	891	1.701	3.68	1.87
130	0.391	668	1.676	5.45	3.74
140	0.416	508	1.652	7.20	5.55
160	0.464	307	1.611	10.5	9.0
180	0.511	195	1.575	13.8	12.4
210	0.582	106	1.530	18.7	17.5
310	0.800	18.5	1.340	33.7	33.2

number of small ones which frequently coalesce later into a few large ones. The mechanical work required to push the water away to form the steam bubbles amounts to about 40 cal/g; the energy going into free surface energy in the bubbles is only about 0.1 cal/g. Thus an estimate can be made of the number of bubbles likely to form. Assume that an average radius of a bubble is $r = 0.2$ cm and its wall thickness is $w = 2 \times 10^{-4}$cm. If the specific volume, ml/g, of steam equals V_{s_0}, then the number, N, of bubbles forming during evaporation of 1 g of water is

$$N = (3/4\pi) \; (V_{s_0}/r^3).$$

The mass, m_b, of bubble walls per gram of steam is

$$m_b = A_b w = (3V_{s_0}/r)w$$

where A_b is the combined area of the bubbles. Substituting the values assumed for r and w, 0.2 cm and 2×10^{-4}, respectively, and the specific volume of 100°C steam, 1674 ml/g, $m_b = 5$ g. Thus, the mass of liquid in the bubble walls is very much larger, five to 10 times that of the steam. Under appropriate conditions it may be expected that practically all of the unevaporated water will be utilized to create the walls of bubbles to form a mist.

Superheated water in a geyser contains more than ample energy to hurl water and steam to great heights at high velocity. Its expansion into steam provides the motive power for doing this. Many geyser jets are thrown to a height of 30 m. This requires the jet to have an initial velocity, v_0, neglecting frictional effects, given by

$$v_0 = (2gh)^{\frac{1}{2}}$$

where g is the acceleration of gravity and h is the height. To throw water 30 m high, v_0 must be equal to at least 2.42×10^3 cm/s. The kinetic energy, KE, per unit mass of such a jet is

$$KE = \tfrac{1}{2} \, v^2 = \tfrac{1}{2}[2.42 \times 10^3]^2 = 2.9 \times 10^6 \text{ ergs/g}$$
$$= 0.29 \text{ joules/g or } 0.07 \text{ cal/g,}$$

a value very small compared to the 40 cal/g of energy used to expand the steam while it is being generated.

The volume occupied by 1 g of erupted fluid is determined mainly by the volume of the steam. For 110°C superheated liquid water changing into steam at 100°C, 1.87 percent transforms into steam generating 31.3 ml/g of steam; for 115°C water, 2.77 percent transforms, making 46.3 ml/g of steam.

The steam generating capacity of the larger geysers is extremely high, exceeding that of many industrial boilers. This is because vaporization occurs not only at the surface of the vent but in the numerous bubbles of steam contained in the body of water. During an eruption of Velikan, for example, the area, bubble surface area plus area of the water exposed to the atmosphere, from which boiling is occurring has been calculated to be about 70,000 m², an area more than 10,000 times that of its 4.5 m² exposed opening.

In its first stage of eruption, a large geyser develops enormous power. Its magnitude is given by

$$\text{Power} = (\text{KE})/\tau$$

where τ is the length of time taken for the superheated water to be discharged from the geyser. Velikan, which takes about 17 s to discharge its 100°C to 115°C, 13.5 metric tons of water, initially develops power at a rate of 180 to 260 kw, generating steam at a rate of 53 to 54 metric tons/hr. The rate of generating mechanical power during an eruption of Old Faithful is between 100 to 200 kw.

3.3 Geysering from a Pool: Fountain or Pool Geysers

The action of a simple fountain or pool geyser is easy to understand qualitatively. An eruption is initiated when deep-lying bodies of hot water abruptly overturn convectively, bringing superheated water up to a shallow level where a portion of it explosively transforms into steam. Before this happens, a certain amount of steady state convection may be established with convection cells and mixing zones developing. Such patterns are easily delineated by throwing bits of thin paper into the pool to be carried along by the convection currents.

Narcissus, Giantess, and Artemisia are classic examples of fountain geysers. Narcissus operates on an alternating two and four hour schedule. Following the eruption which is preceded by the two hour interval, the basin is completely emptied by suction through fissures in the bottom of the reservoir. It then begins to slowly fill with hot water, becoming completely full after about three hours. Temperature of the water (Fig. 2-7) in the lower chamber gradually increases during this time whereas the water in the wide-mouthed basin slowly cools, presumably by evaporation. Every hour or so there is an excursion of hot water upward accompanied by downward movement of cooler water. Finally, one of the rising blobs of superheated water contains enough excess heat energy to remain superheated after its encounter with the cooler upper basin water, exploding into steam as it approaches the surface and violently agitating the water in both chambers. Blob after blob of superheated water begins to move upward giving rise to a succession of explosions, which, after a few moments, erase all temperature and density gradients. The explosions cease and the eruption is over.

Following the four hour interval, the reservoir almost but not quite empties itself through bottom fissures after the eruption is finished. The next eruption takes place after only a two hour interval, indeed even before the upper basin is completely full of water. During this filling, the water temperature rises faster. This happens in other fountain geysers, notably Kotegu in Kamchatka.

Both Giantess and Artemisia in Upper Geyser Basin have surface basins of considerable size and depth filled with water that during an eruption domes up, breaking into a series of violent splashes. Giantess's basin is elliptical, 8 by 10 m in diameter and about 8 m deep and sits in a mound which slopes in a staircase-

like terrace toward the bank of the Firehole River. Artemisia's basin is circular, 15 m in diameter and 14 m deep.

In the power of its eruption, Giantess far exceeds Artemisia even though the latter's basin is much larger. At the time of an eruption, a great mass of hot water is hurled 18 m in the air with higher spurts shooting up to heights of 75 m. After about 36 hr these stop, to be followed by a powerful steam phase lasting 12 hr. Much more water is discharged during the eruptive bursts of Artemisia but less violently. The water in its pool suddenly domes up, bursting into jets shooting in every direction but with none reaching heights greater than 10 m. Play is very regular; it lasts for nearly a half hour with about a 12 hr quiet period.

The Great Geysir, though now inactive, also functioned as a fountain geyser. The discernible part of its plumbing consisted of a 3 m diameter, 20 m deep tube topped by a 16 m diameter 1 m deep basin. It took 10 to 12 hr to refill after an eruption. An eruption was heralded by the water doming at the surface. The initiation appeared to be located at about the 10 m depth.

Round Geyser is a 6 m deep cylindrical shaft about 1 m in diameter. For several hours before an eruption, its waters circulate in a well developed single convection cell, water rising at the center and descending at the rim, with the bottom temperature staying between 115°C and 120°C. Suddenly for no observable cause, the water begins to flow over the edge, followed in a few seconds by almost instantaneous vaporization of all the water in the reservoir, throwing steam and spray in one big burst to a height of 20 to 30 m.

3.4 Geysering from a Pipe: Columnar or Cone Geysers

A majority of columnar or cone geysers erupt from fairly small vertical or nearly vertical, fairly uniform tubes, usually ejecting a single steady jet of hot water and steam, emptying all or most of the tube in a few minutes. The water supply in the simplest of these is a steady flow of superheated water into the bottom of the tube, up it, and finally out through the surface. During the water's upward course, as it fills the tube after an eruption, it cools and soon there develops a column of water cooler at its top than at its bottom. As hot water continues to flow, a few steam bubbles can be expected to develop at those locations in the column where the temperature of the upward flowing water is higher than the ambient boiling point. The bubbles will ascend rapidly at first but will then be cooled in the overlying water, often collapsing and disappearing. Eventually a metastable energy state develops within the tube which is relieved by the ensuing eruption.

Experiments on models of such a geyser indicate that it will erupt only if there is a constriction in the geyser tube where rising steam bubbles are caught. The eruption seems to be triggered when all of the water below the constriction has been heated to the boiling point with no water cool enough to condense the steam bubbles. Bubbles entering the constriction become trapped and lift the overlying cooler water up and out of the geyser as a mass, precipitating an eruption by

reducing the hydrostatic pressure on the lower-lying hotter waters. Further, constriction may lead to the current becoming swifter and the evolution of steam bubbles more rapid.

Temperature measurements corroborate the applicability of the model. Eruptions were found to be initiated at a relatively shallow depth by steam bubbles powerful enough to push out a plug of overlying water. Sometimes it is necessary for the overlying water to be pushed up and out in increments. At Old Faithful, as many as 20 large splashes occur before an eruption gets underway. The temperature-depth curves in all the geysers show an upper regime of cooler water separated by a mixing zone, apparently at a constriction, from deeper, hotter water.

In models without constrictions, the steam bubbles rise into the upper cooler water where they collapse at first but as the water becomes hotter, get all the way to the surface where steady boiling commences. The action of these models corresponds to the behavior observed in natural boiling hot springs.

By making a number of fairly realistic assumptions, it is feasible to carry through an analytical analysis of the hydraulic and thermal regimes of a simple columnar geyser and to predict in considerable detail certain features of its behavior. The analysis shows that two stages of equilibrium develop. First, the geyser tube fills with hot water before boiling starts; and second, boiling begins within the tube with the level at which it takes place moving progressively downward during the eruption, reaching some final level by the end of the eruption. This final level is usually very much deeper down in the tube than the point at which vigorous boiling started. The greater the difference between these two levels the greater the tendency of the geyser to erupt.

Assume that the geyser tube is vertical and uniform in cross section and extending so deep that it can be considered infinitely long. Assume further that V_w milliliters of T_1 degree absolute water enters per square centimeter of cross section of the tube per second and that T_1 is greater than sea level boiling point, 100°C or 373°K. Assume also that in the beginning, the geyser tube is filled with water whose temperature is constant, T_1, from the bottom of the tube up to a depth, d_1, the depth at which the hydrostatic pressure has decreased to the point where the T_1 degree water will boil. The temperature of the water is assumed to decrease from d_1 on up to the surface in such a way that at nowhere in the tube is the vapor pressure of the water high enough to overcome the ambient hydrostatic pressure and have boiling occur. The upward flow of water will eventually lead to local boiling regimes within the tube. Such boiling episodes may occur a number of times before an eruption starts but one will finally precipitate it.

At any depth, boiling will vaporize m_s of each m_w g of the ascending water where

$$m_s = m_w(T_1 - T)\,\sigma. \tag{3-2}$$

T is the temperature of the residual water after boiling has occurred and σ is the heat of vaporization of the water, 540 cal/g. The much lighter, newly-formed steam displaces a portion of the denser water, reducing the hydrostatic pressure,

usually allowing boiling to take place at depths greater than d_1 unless the flow of water into the geyser tube is exceptionally high. The magnitude of this effect can be calculated.

The volume of steam, V_s, formed from the m_s g of vaporized water is

$$V_s = m_s/\rho_s = m_w(T_1 - T)/\sigma\rho_s \tag{3-3}$$

where ρ_s is the density of the steam in the bubbles. Thus the approximate average density $\bar\rho$ of the column of fluid at depth becomes

$$\bar\rho = m_w/(m_w + V_s) = 1/\left[1 + (T_1 - T)\sigma\rho_s\right] \tag{3-4}$$

since the density of water is nearly equal to one.

Now the gradient of the hydrostatic pressure is given by

$$dp/dd = \bar\rho(d) \tag{3-5}$$

which can be rewritten as

$$(dT/dp)(dd/dT) = 1/\bar\rho \tag{3-6}$$

where p is the vapor pressure of water at depth d.

Substituting from Eq. (3-4)

$$dd = \left\{(dp/dT) + \left[(T_1 - T)/\sigma\rho_s\right](dp/dT)\right\} dT. \tag{3-7}$$

On integration from the surface, located at depth d_1, at which boiling commenced, where the temperature is assumed to be just at boiling, 373°K, to an equilibrium depth, d_e, down to the which boiling now extends as a consequence of the reduction in hydrostatic pressure resulting from presence of steam bubbles in the overlying waters, gives

$$d_e = p_{T_1} - p_{373} + \int_{T = 373}^{T_1} \left[(T_1 - T)/\sigma\rho_s\right](dp/dT)dT. \tag{3-8}$$

Using as units of pressure centimeters of water makes

$$p_{T_1} - p_{373} = d_1 \tag{3-9}$$

provided the water lying above the depth, d_1, is too cold to boil.

The integral in Eq. (3-8) can be evaluated by means of the empirical relationship

$$(1/\sigma\rho_s)(dp/dT) = 1.145 - 0.009t - 0.00004t^2 \tag{3-10}$$

where t is the temperature in °C and the pressure in the steam bubbles is equal to the ambient hydrostatic pressure. For convenience, setting Δd_1 equal to the value of the integral, then

$$\Delta d_1 = \int_{t = 0}^{t_1} (t_1 - t)(1.145 - 0.0019t - 0.00004t^2)dt \tag{3-11}$$

where

$$t_1 = T_1 - 373.$$

From Eqs. (3-8) and (3-9), it is seen that

$$d_e = d_1 + \Delta d_1. \tag{3-12}$$

This equation implies on the basis of the assumptions made that if boiling starts at depth d_1, it will propagate deeper down, finally reaching another equilibrium at the greater depth $(d_1 + \Delta d_1)$. The value of Δd_1 is strongly temperature dependent, becoming quite large for water temperatures much above 100°C. Some computed values of Δd_1, the distance from the first to the new equilibrium position, are listed in Table 3-2.

Having estimated values of Δd_1, it is possible to estimate the duration of play of a columnar geyser as a function of water temperature but first it is instructive to develop a relationship between water temperature and height of play.

The rate of boiling governs the height of the eruption. To reach a height, h, the water droplets must leave the geyser orifice with the velocity, v_o, given by

$$v_o \geq (2gh)^{1/2} \tag{3-13}$$

where g is the acceleration of gravity, 980 cm/s². The volume of fluid, steam and water, passing through each square centimeter of orifice area per second is also numerically equal to v_o or

$$v_o = V_s + V_w \tag{3-14}$$

where V_s and V_w are the volumes of steam and water, respectively. The volume of steam in the ejected mixture of fluids is given by Eq. (3-3). According to Table 3-1, the density of steam at atmospheric pressure is 1/1674 g/ml. Since $\sigma = 540$ cal/g, V_s in the same units is given by

$$V_s = 3.1 \, t_1 \, m_w = 3.1 \, t_1 \, V_w \tag{3-15}$$

where t_1 is the difference between the ambient boiling point at the orifice and the temperature of water deep within the tube.

The ratio of volume of steam to volume of water exiting from the orifice is therefore approximately equal to $3t_1$. This result indicates that whereas at geyser temperatures ranging from 110°C to 120°C ($10° < t_1 < 20°$) only 2 to 4 percent by mass of the water is transformed into steam (Table 3-1) and the volume of the ejected steam will be 30 to 60 times that of the ejected water droplets.

Substituting the above results in Eq. (3-14) gives

$$q_w = (2gh)^{1/2}/(3t_1 + 1) \tag{3-16}$$

where q_w is the rate of flow of water just deep enough in the geyser that no steam

Table 3-2. Computed Values of Δd_1 for Several Geyser Water Temperatures (Adapted from Thorkelsson, 1940)

Temperature of Geyser Water (°C)	d_1 (m)	Δd_1 (m)
101	0.4	0.6
110	4.3	56.9
120	9.8	226.0
130	17.1	505.0

has yet formed, the temperature here being $(373 + t_1)°K$. Velocities required for various temperature waters to maintain a continuous 50 m high eruption are listed in Table 3-3.

Normally the upward flow velocity is not this high so that the water level in the tube drops. When there is no upward flow, this boiling completely empties the tube. The times required to empty a 20 m long tube under these conditions are listed for several temperatures in the same table. If there is some but not totally adequate upward flow to maintain a steady condition, then the velocity, $dd/d\tau$, with which the water level falls is given by

$$dd/d\tau = q_w - V_s \qquad (3-17)$$

where τ is time.

Substituting for q_w in Eq. (3-16) and solving for h gives

$$h = \Delta d_1 - \Delta d = 0.57 \, t_1^2 - \Delta d. \qquad (3-18)$$

According to this equation, the geyser eruption will be the highest at the beginning when $\Delta d = 0$, subsiding gradually as the lower waters participate more and more in the eruption and Δd increases. It takes some time at the beginning for an eruption to get up to full speed since the cooler overlying waters must first be thrown out. Neglecting this effect, the maximum, equal to $0.57 \, t_1^2$, therefore varies as the square of the number of degrees of superheat contained in the waters lying a few tens of meters below the geyser surface opening. Water at 110°C is unable to produce an eruption greater than 57 m high.

The eruption will stop when equilibrium is reached at the depth $(d_1 + \Delta d_1)$. The time required to reach this state, τ_1, is the length of play of the geyser which can be theoretically calculated. In order to develop a 50 m high eruption with 120°C water, the water must flow up the tube at a velocity of 51 cm/s. Suppose now that the natural rate of inflow is 1 cm/s. Then the velocity at which the boiling region moves downward is 50 cm/s, requiring

$$\tau_1 = \Delta d_1/50 \text{ s} \qquad (3-19)$$

to descend from the d_1 level to the $(d_1 + \Delta d_1)$ level. Using Table 3-2, $\Delta d_1 = 226$ m for 120°C water, giving, when substituted in Eq. (3-19) 7 min 32 s for the length of play, a duration of reasonable magnitude. After this time, the eruption

Table 3-3. Water Velocities Needed to Maintain a 50 m High Eruption and Times to Empty a 20 m Long Tube Assuming No Upward Flow (Adapted from Thorkelsson, 1940)

Water Temperature (°C)	Velocity (m/s)	Time to Empty Tube (s)
120	0.51	40
125	0.41	50
130	0.34	60

ceases for lack of hot water and a new eruption could not begin until the tube refilled and the temperature at 9.8 m (see Table 3-2) had reached 120°C.

The duration of play is seen to depend on both the temperature and rate of influx of water. When there is no influx of water, τ_1 is given by

$$\tau_1 = 2(1 + 3.1\ t_1)(\Delta d_1/2g)^{1/2} \tag{3-20}$$

which predicts a 1 min 50 s play if the water is at 110°C and a 7 min 10 s play if at 120°C.

The above calculations, although highly idealized, provide a great deal of insight into why and how certain kinds of geysers operate as they do. Factors that will influence the direct applicability to some extent are:

• the assumption that steam is weightless and has no heat capacity other than its heat of vaporization. This does not affect the results appreciably since such a small proportion of the mass involved is steam;

• the assumption that there are no cooling effects of the geyser tube. In the beginning, they may be relatively quite large but at the time of the eruption they become relatively unimportant. The most serious situation is where the top end of the geyser tube widens out into a basin and surface cooling becomes pronounced, sometimes severely affecting the developing of geyser action; and

• the assumption that the acceleration of water up the tube during an eruption will increase the pressure exerted against the lower waters. An analysis of this effect shows that the rate of propagation of the boiling down the tube is slowed down by the acceleration given the water and cannot exceed a certain limit.

Calculated values of the velocity at which the boiling surface moves downward are listed in Table 3-4 from which it is evident that the maximum rate of boiling is almost constant, 1.05 m/s in the temperature interval 110°C to 130°C. However the rate decreases as the boiling progresses downward and Δd increases. It is obvious from the table that a very high rate of flow of water up the tube, greater than 1 m/s, would be required to prevent boiling from propagating downward. Such a condition would cause a perpetual spouter to develop.

Friction in geyser conduits has not been taken into account. It can also be important in regulating geyser action. Frictional resistance should remain nearly constant all the way up the tube since while it increases near the top due to the higher speed of exiting fluids, it will decrease at the same time by virtue of the

Table 3-4. Calculated Rates of Boiling in Geyser Tube (Adapted from Thorkelsson, 1940)

Water Temperature (°C)	Velocity at Which Boiling Surface Moves Downward (m/s)
110	1.03
120	1.05
130	1.05

fact that more bubbles form there. The net effect of frictional resistance is to suppress boiling since it is a force acting in the same direction as hydrostatic pressure. The effect was well known by Icelandic farmers who prevented certain annoying geysers from erupting by filling their tubes with rocks. Contrariwise, many geysers whose tubes have become partially filled and become constant spouters can be returned to their normal action by the removal of the debris. Or the action may only be somewhat modified as was the case with Littli Geysir in Iceland whose vent is enclosed by stone, probably man-placed. During its frequent periodic eruptions, water and steam shoot forth in all directions from among the rocks. Some jets reach heights of 10 to 13 m. An eruption is presaged by initiation of subterranean splashing and a gradual increase in the amount of steam discharged. The eruption itself builds up slowly, reaching a maximum in about 10 min, after which the water spouts begin to die down and the geyser becomes quiet in about another 10 min.

It has been tacitly assumed in the preceding discussion that the liquid and vapor phases of the geyser waters are uniformly mixed and move together through the system exerting a hydrostatic pressure as if the density of the mixture were inversely proportional to the combined column of steam and water. This is essentially true although near the mouth of the geyser, the bubbles coalesce and form larger bubbles that ascend at a higher velocity than the water.

Another factor not taken into account in the above derivations is variation of the rate of influx of water, due especially to variation in hydrostatic pressure in the geyser tube. Such variation plays an important role in the rate of heating of a buried reservoir. A simplified analytic treatment of this effect indicates that water will flow in more rapidly under low hydrostatic pressure than high. Increasing the rate of water flow causes the boiling to propagate more slowly downward. As a consequence of the slowing down of the rate of propagation of the boiling surface, the geyser eruption lasts longer and produces higher and more frequent bursts than it would when the water is coming in less fast. All of the equations discussed above show fairly well that geyser eruptions of substantial height cannot continue for long periods unless especially large amounts of hot water are supplied from other subterranean channels and cavities serving as auxiliary reservoirs.

Although an eruption could possibly be terminated in a number of ways, it seems most likely that is terminated because the system has either expended all of its excess heat, which is what generally happens in a fountain geyser, or it has thrown out all of its excess hot water, the usual situation in a columnar geyser.

Maximum pressures exist throughout the plumbing system of a columnar geyser just prior to an eruption and as the eruption proceeds, the pressure within decreases. With the development of substantial pressure differentials between the geyser's reservoirs and the surrounding rock mass, the water in the saturated pores and fissures of the rock mass tends to drain into the reservoir. A zone of falling water levels forms around the tube. The scarcity of water, especially in the rock mass which still contains excessive heat, is conducive to the develop-

ment of vast quantities of steam. This steam first forces all of the remaining liquid water out of the geyser tube and then roars forth to develop the powerful terminal steam phase characteristic of the behavior of most columnar geysers.

3.5 Complex Geyser Systems

Simple columnar and pool geysers can only generate a single thrust of water and steam, or a single series of steam bubbles. More elaborate systems must be invoked to account for the complex behavior of many geysers, especially for the large amounts of water thrown out by some geysers, and for the multiple thrusts seen at Grand Geyser, Great Fountain, and others.

A fairly simple idealized model to analyze, to account for large volumes of discharged water is the one shown in Fig. 3-7. It consists of a standpipe fed from its base by a single large reservoir. There is a fluid heat source entering through channels, a; a reservoir, A; a supply of cold water entering directly into the reservoir through channel, b; and an orifice, c, through which the geyser discharges. The volume of the tube, C, is assumed to be negligible compared to that of the reservoir. These elements interact to produce an eruption, occurring explo-

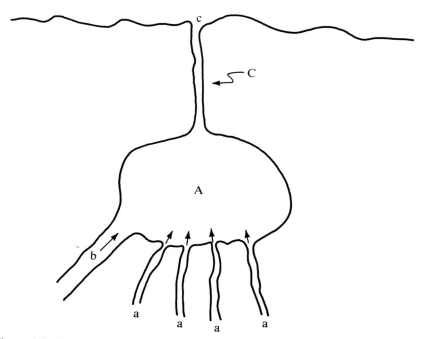

Figure 3-7. Model of complex columnar geyser. (Suggested by Merzhanov et al., 1975, private communication.)

sively as a result of rapid conversion of the excess heat energy, W, of superheated water which amounts to

$$W = c\Delta T \tag{3-21}$$

where c is the specific heat of the water and ΔT is the difference in temperature of the boiling water under the ambient atmospheric pressure, p_o, and the temperature of the boiling point of water under the pressure, p, corresponding to the depth, d, at which the superheated water resides. p is given by

$$p = p_o + \rho g d \tag{3-22}$$

where ρ is the density of the water, taken here to be equal to 1 g/ml. Eq. (3-22) indicates that the higher the value of $\rho g d$, the greater is the energy of superheating and the more violent the eruption.

The first phase following an eruption is the refilling of reservoirs and tubes and the heating of their newly accumulated water. The latter is accomplished by an influx of small quantities of hot, heat-carrying water or steam that rapidly mixes with the cooler water. Some heating is accomplished by heat conduction through the walls of the reservoir but the amount is considered negligible. The rate of flow of the hot water which comes from great depth, is nearly independent of the pressure in the reservoir whereas that of the cold water of shallow origin may vary somewhat, being affected by the difference in pressure between that in the water-bearing bed and that in the reservoir. The heat source may be steam which after condensation is equivalent to water at temperature, T_h, where

$$T_h = T_s + (\sigma/c) \tag{3-23}$$

where T_s is the temperature of the steam.

It is assumed that the volume, V_A, of the reservoir is much larger than that of the exit channel, V_c, and that any pressure gradients in the reservoir are negligible.

As the reservoir begins to fill, all residual water is at temperature, T_o. As time goes on, the amount of water in the reservoir, which is a mixture of hot and "cold" water, increases at a steady rate governed by the differential equation

$$\rho(dV/d\tau) = q_1 \tag{3-24}$$

and its temperature is constantly changing with mixing in accordance with the equation

$$\rho V(dT/d\tau) = q_h(T_h - T) + q_c (T_c - T) \tag{3-25}$$

with $\tau = 0$, $V = V_o$, and $T = T_o$ at the completion of an eruption. The quantity, q_1, is equal to the total rate of inflow of water ($q_h + q_c$), q_h and q_c being, respectively, the rate of influx of hot water or its equivalent in steam, and cold water.

Solving the differential equation yields

$$V = V_o + (q_1/\rho) \tau \tag{3-26}$$

and

$$T = T_{e_1} + (T_1 + T_{e_1}) \left[1 + (q_1/\rho V_o)\tau\right]^{-1} \qquad (3\text{-}27)$$

where

$$T_{e_1} = (q_c T_c + q_h T_h)/q_1 \qquad (3\text{-}28)$$

is the equilibrium temperature after mixing. Note that until the reservoir is completely filled, the temperature of its water will decrease in the most common situation where more cold water than hot enters it.

The condition of most relevance to geyser activity is when T_{e_1} becomes greater than T_o. The time, τ_1, to fill the reservoir is given by

$$\tau_1 = (V_A - V_o)(\rho/q_1). \qquad (3\text{-}29)$$

The temperature, T_1, at that time is

$$T_1 = T_{e_1} + (V_o/V_A)(T_o - T_{e_1}). \qquad (3\text{-}30)$$

Once the reservoir is full, the inflowing waters begin to fill the exit channel, the rate of rise of the water level in the channel depending upon the rate of inflow of water and the cross sectional area of the channel. The volume of water in the reservoir, of course, remains constant but the pressure there increases due to the weight of the overlying water in the channel. This increased pressure reduces the pressure difference between that of the cold water in the rock mass from which it enters, and the water in the reservoir, usually decreasing the rate of influx of water. The amount of decrease can be approximately determined from the application of Darcy's law which relates permeability to applied pressure. The effect is appreciable for porous material, gravels, and heavily fractured rocks and can be taken into account although the calculations become more involved. For dense competent rock masses, it can be neglected.

For the present, neglecting any variation in flow rate, the time, τ_2, after an eruption when the channel becomes full, assuming that it is of length, D, and uniform cross section, A, is

$$\tau_2 = \tau_1 + AD/q. \qquad (3\text{-}31)$$

The temperature, T_2, is assumed uniform throughout the reservoir and channel at just the time that the channel becomes completely filled. From then on, the pressure stays steady and the total amount of water in the system, reservoir and channel, remains constant.

At the same time the water begins to heat up, eventually precipitating an eruption. It is assumed here that the eruption will occur when T_{e_2}, the equilibrium temperature achieved by mixing of the hot and cold fluids in the reservoir, becomes greater than the boiling temperature, T_1, corresponding to the pressure, p_2, in the reservoir. At this time, steam bubbles form, forcing the water out of the channel, relieving the pressure on the reservoir, and initiating violent boiling. Solving the heat balance differential equation

$$V_A = (dT/dt) = q_h(T_h - T) + q_c(T_c - T) \qquad (3\text{-}32)$$

gives an expression for determining the temperature T at any time:

$$T = T_{e_2} - (T_{e_2} - T_2) \exp\left[-(q_1/V_A)(\tau - \tau_2)\right]. \qquad (3\text{-}33)$$

The equilibrium temperature as before is given by

$$T_{e_2} = (q_c T_c + q_h T_h)/q_1. \qquad (3\text{-}34)$$

Any effect of pressure on rate of flow of the cold or hot water would change the values of q_c and hence q_1. In fact, calculations on Velikan's thermal and hydrologic behavior indicate that the reduction in flow might be as much as a factor of 10.

The interval of time, τ_I, from the end of one eruption to the beginning of the next will be the sum of the time to fill the reservoir, plus that to fill the channel, plus that to further heat the water in the reservoir up to its boiling. This total time, not taking into account reduction in flow of cold water, is then

$$\tau_I = \rho(V_A - V_0)/q_1 + (\rho V_A/q_1) \ln\left[(T_{e_2} - T_2)/(T_{e_2} - T_I)\right]. \qquad (3\text{-}35)$$

A full analysis of the model using the above equations has led to the following conclusions:

If $T_{e_1} < T_I$ and if $\begin{cases} T_{e_2} > T_I \\ T_I < T_{e_2} < T_I \quad \text{it is a} \\ T_{e_2} < T_I \end{cases}$ $\begin{cases} \text{geyser} \\ \text{boiling spring} \\ \text{hot spring.} \end{cases}$

If $T_{e_1} > T_I$ and if $\begin{cases} c(T_{e_1} - T_I) < \sigma & \text{it is a fumarole} \\ \\ \begin{cases} T_{e_2} > T_I \\ T_{e_2} < T_I \end{cases} \begin{cases} \text{geyser or} \\ \text{pulsating spring} \\ \text{boiling spring.} \end{cases} \end{cases}$

The results of one numerical calculation are plotted in Fig. 3-8. The geyser parameters used are listed in Table 3-5. These are believed to be reasonably characteristic of Velikan. The total water discharged during one eruption cycle indicates that in addition to its large vent, it must also have a cavity or cavities for storing water, the volume of which exceeds by several times that of the vent. Velikan erupts every 3 to 4 hr, throwing a column of water to a height ranging from 30 to 50 m for about 90 s. This discharge of hot water is followed by 20 to 30 minutes of powerful steam ejection. The vent, 6 m deep with a 4.5 m² opening, has a storage capacity of 13.5 metric tons of water. Calculations of the sort discussed in Sec. 3.4 show that the vent would empty itself of water in about 20 s; however, the jet continues playing for more than another minute, clearly indicating that additional water is stored in subterranean cavities. Let the total eruption time be 90 s, and assume that the intensity of the last part of the eruption is only half that of the initial state. Then the additional erupted mass of steam and water, m_1, is

$$m_1 = (m/2)\left[(90 - \tau)/\tau\right] = 1.30 \text{ m to } 2.15 \text{ m} \qquad (3\text{-}36)$$

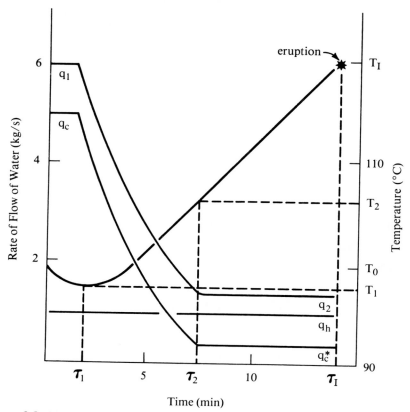

Figure 3-8. Results of sample calculation of the temperature and hydrologic regimes existing in the model geyser of Fig. 3-7. (Courtesy of Steinberg et al., private communication.)

where m is the mass of water contained in the vent just before an eruption. The lower value of m_1 applies to 115°C temperature water and the higher figure to 110°C water. The mass, m_s, discharged as steam during the subsequent 20 min play is

$$m_s = x \, m \, \tau_s / 6 \, \tau \qquad (3\text{-}37)$$

assuming that the rate of discharge of water has not dropped to one-third of its earlier value. τ_s is the length of the steam phase and x is the fraction of water changed to steam, assumed equal to 0.167. Substituting values, then

$$m_s = 0.22 \, m.$$

The mass, m_b, of water forming bubble walls must also be taken into account. From Table 3-1, this is given by

$$m_b = 6 \, m_s.$$

Table 3-5. Values Used to Obtain Temperature and Hydrologic Regimes Plotted in Fig. 3-8 (Adapted from Steinberg, Personal Communication)

V_A	$= 2 \times 10^3$ liters
V_o	$= 1.4 \times 10^3$ liters
V_c	$= 10^3$ liters
D	$= 10$ m
q_h	$= 1$ kg/s
q_c	$= 5$ kg/s
$q_c{}^*$	$= 0.5$ kg/s
T_e	$= 200°C$
T_c	$= 75°C$
T_1	$= 100°C$
T_2	$= 120°C$

$q_c{}^*$: Flow rate of cold water when geyser reservoir and channel are full.

Thus the total mass, m_T, discharged during the eruption is

$$m_T = m + m_1 + m_b \tag{3-38}$$

or

$$m_T = 3.62 \text{ m to } 4.47 \text{ m,}$$

the lower amount corresponding to 115°C, and the higher to 110°C temperature water. Since m is the 13.5 metric ton mass of water discharged from the vent, then m_T ranges from 48 to 61 metric tons. The subterranean cavity or cavities within Velikan thus must hold three to five times as much water as the vent.

Other qualitative details of Valikan's behavior support this conclusion: one is that Velikan's vent remains empty for some time after an eruption, finally slowly filling in about 1.5 hr. Assuming that water is entering far underground, such behavior is completely consistent with the notion that at least two subterranean cavities must be filled before water will begin to enter the vent. After the vent is full, water begins to overflow the rim of the basin indicating a constant influx of water. Also, the water level in the basin pulsates up and down in a sort of double and at times triple rhythm, accompanied by periods of local boiling and overflow.

The Atami Geyser is believed to have had a somewhat more complex geyser system than Velikan. It is located very close to the sea coast and was quite active until 1924 when it ceased to erupt, probably as a result of human intervention. Its normal eruption, occurring about every 5 hr, usually consisted of five successive alternate projections of hot water and steam. As the time of the eruption approached, subterranean rumblings were heard, with boiling water intermittently appearing just inside the mouth every 1 to 2 min. After about 45 min of this kind of activity, small quantities of water intermittently flowed out of the mouth, the

amount of water and steam, especially steam, increasing with each surge until finally almost entirely steam was roaring out of the opening. Suddenly the flow of steam stopped, to be followed shortly by a tremendous gush of water. After about five such gushes, each interrupted by violent steam flow, with the entire process taking about 2 hr, the geyser became quiet and remained so for about 3 hr with steam lazily rising from its mouth.

Occasionally, Atami exhibited abnormal behavior. The regular sequence of periodic eruptions was interrupted by an extraordinarily long eruption called a *nagawaki*, lasting 12 hr, which was followed by a fairly long, approximately 7 hr, quiet period before the regular sequence of eruptions was resumed. The Japanese refer to such a performance as an *oyakobuke*, meaning mother and daughter eruption.

Extensive studies both in the field and on laboratory models establish rather conclusively that Atami had two large underground storage reservoirs, illustrated in cross section in Fig. 3-9. Cavity, A, which lies at considerable depth, receives water through channel, b, and discharges it through channel, a. A nearby cavity, C, is fed water through channel, d, and discharges through channel, c, into channel, a. It is assumed the eruption is initiated in cavity, A, where the temperature of the water reached a critical value before that in cavity, C. After a certain amount of water and steam have been discharged, the pressure in the exit channel, a, falls low enough that water enters channel, a, and cavity, A, via channel, c, quenching the eruption momentarily. Pressure then again builds up producing

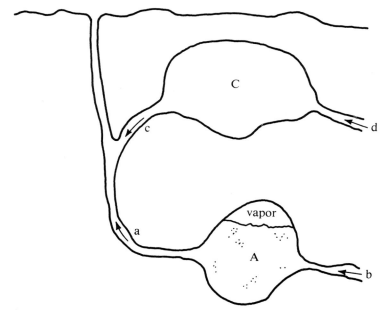

Figure 3-9. Inferred cross section of Atami Geyser. (Adapted from Honda and Terada, 1906.)

a second gush of water from A. After several such interactions all of the excess energy has been expended and activity ceases. The nagawaki undoubtedly originates at considerable depth and is not affected much by the shallower parts of the system.

Model experiments made with somewhat similar systems shown in Fig. 3-10 produce eruptions exhibiting two and sometimes three thrusts. In the experiments, the two reservoirs are filled with water to a level above the connection between the two. Both are heated from below and the actions in the two reservoirs are strongly coupled. The reservoirs cannot ever boil vigorously at the same time since that would create an unstable condition. Assume that the rate of boiling in one reservoir, B, increases slightly. Then the level of the water located above the junction, a, rises, increasing the hydrostatic pressure and hence the temperature which water must attain in the other reservoir, A, to boil. Depression of the rate of boiling in A decreases the number of bubbles in it. Whereupon water flows into A, decreasing the hydrostatic pressure in B, and increasing the rate of boiling in it. Thus, boiling moves from one chamber to the other and water should flow back and forth between the two as first one boils vigorously and then the other. Indeed, such strong oscillatory flow was observed in the experiments by watching the dust and bubbles carried along by the changing currents. Many complete oscillations occur before an eruption. The oscillations also generate tremors resulting from the mechanical forces associated with the water flow.

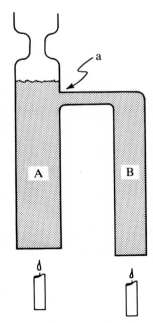

Figure 3-10. Laboratory model corresponding to Fig. 3-9. (After Anderson et al., 1978.)

During an eruption of such a system and even more complex ones, the reservoirs discharge in series. As the first throws water and steam from the orifice, some water also flows into the other reservoirs which increases the pressure and hence suppresses the tendency to boil in them. Another reservoir then starts its eruption, during which it may force enough water back into the first reservoir or other reservoirs to enable them to erupt even a second time.

While the action of simple fountain geysers is understood in a general way, it is almost impossible to specify theoretically and quantitatively the hydrologic and thermal regimes involved in the eruptions of more complex types since very little is known of their individual plumbing systems. Certainly the requisite regimes are strongly dependent on the geometric configurations of the geyser's reservoirs, the source and location of heat and water and most important, the delicate interactions among these.

CHAPTER 4

The Role of Gases in Geysers

4.1 Formation and Evolution of Gases

Gas, especially steam, is the primary agent in precipitating and sustaining geyser activity. Thus far only the presence of steam has been taken into account. All geyser waters contain other gases, occasionally in substantial amounts. Some of the relative amounts of noncondensible gases in geysers and hot springs are listed in Table 4-1. Carbon dioxide, by far the most abundant, ranges from about 80 percent to almost 100. Other gases, oxygen, carbon monoxide, hydrogen, methane, nitrogen, argon, and hydrogen sulfide occur only in minor amounts. The noncondensible gases as well as some of the steam are almost entirely of volcanic origin except perhaps for the nitrogen and argon, part of which are derived from the atmosphere. The composition of gases in those spring waters located in fresh, hot volcanic areas is generally considerably different. In the Kamchatka spring waters and fumarole gases, carbon monoxide is more abundant than carbon dioxide, especially in those fumaroles with temperatures above 150°C. The amount of hydrogen, which ranges from nil in some locations to more than 40 percent in others, is also more plentiful in these basaltic volcano areas that exhibit high proportions of carbon monoxide. In a few locations, odoriferous hydrogen sulfide constitutes as much as 20 percent of the active gases; occasionally, sulfur dioxide is found to exceed 20 percent of the total active gases.

The presence and percentage of the active gases varies from spring to spring, being more dependent upon the type of spring than on geographic location. The alkaline springs, especially the larger ones, and the main geysers are, as a rule, conspicuously high in nitrogen and methane. Daisy Geyser contains almost 18 percent nitrogen and 1.6 percent methane. Acid and sulfate springs give off relatively large amounts of gas whereas alkaline waters are relatively deficient in

78

Table 4-1. Noncondensible Gases Effusing from Hot Springs; wt % of Total (Adapted from Allen and Day, 1935; Barth, 1950)

	CO_2	O_2	H_2	CH_4	H_2+A	H_2S
Yellowstone						
Black Growler	97	0	0.4	0.1	0.5	2.5
No. 70, Norris Basin	89	1.4	2.3	0.6	6.6	0
Chinaman, Upper Basin	92	0	0.5	0.8	7.2	0
Daisy, Upper Basin	79	1.9	0	1.6	17.9	—
Drillhole, Upper Basin	95	0	0.1	0.1	4.2	0.6
Midway Basin	91	0.3	0	0.2	9.0	0
Imperial Geyser, Lower Geyser Basin	86	2.0	0	0.4	11.8	0
Heart Lake Basin	88	0.8	0	2.4	8.6	0
Crater Hills	98	0.1	0	0.1	1.8	0
Iceland						
Geysir Group	77	1.7	0	—	21.3	0.2
No. 169, Reykir, Olfus	61	0.4	—	2.3	16.1	12.0
No. 217d, Reykjanes	91	1.1	0.5	n.d.	5.3	2.1
Mvatn	74	—	9.3	—	3.3	13.9

gas content. These latter contain substantial amounts of bicarbonates presumably because the carbon dioxide originally present has combined with the minerals to form bicarbonates.

The gas escape rate ranges widely even in springs of the same type; however the rate in acid and sulfate areas is about 10 times as great as in alkaline areas. Substantial variations in the amount of gas discharge in similar ground are probably controlled by differences in the fracture patterns of the underlying rock mass. Alkaline areas discharge essentially no hydrogen sulfide so that such areas are completely free of its distinctive and unpleasant odor.

It is not feasible to measure the total amount of gas which escapes from a spring. It is only practical to make a qualitative comparison of gas discharged from one spring with that of another. This is done by observing the time required to collect a given amount of gas in the most active regions of each. Using the same equipment, a 10 cm diameter funnel, 500 ml of gas was about the maximum that could be collected from any one spring in one day in the Upper and Lower Geyser Basins. Giantess yields about 1 ml per hour; Great Fountain, a little less; and Riverside, essentially none at all. On the other hand, Daisy provided 100 ml of gas in about 7 min; Artemisia, in 30 min; and Oblong in 110 min.

The discharge of gases in a few springs is periodic. Such action is sometimes referred to as a subterranean eruption but even quite violent activity is damped before it reaches the surface by collapse of the bubbles in the cooler basin water.

Much of the gas originally dissolved in the liquid magma and given off when

it freezes becomes trapped in the various types of rocks that form the magma. Since magma contains a high percentage of water, the amount of water in igneous rocks is more than that of all the other trapped gases combined. It is the highest in glassy rocks which are the least altered.

Rhyolite pumice contains more gas exclusive of water than obsidian. Obsidians exhibit the greatest range in total gas content primarily due to absorption of meteoric water; the water content will range from 90 to 99 percent of the total gases. Although gas makes up only a small percentage of the rock by weight, by volume it can be quite large. The gas from the average granite, when heated to 100°C under pressure of one atmosphere, will occupy a volume 30 times as great as the rock from which it comes, assuming about 2 percent water content by weight. In the case of an average basalt, the volume of gas will be 87 times as great. Other significant findings relating to the gas content of in situ rocks are that the total gases including water from unaltered rocks range, when converted to steam, from about 1 to 200 ml/g of rock; that total gases other than water are less, ranging from 0.14 to 6 ml/g of rock; that the total gas content in geologically recently emplaced andesites and basalts is about 6 ml/g of rock, with water constituting approximately 80 to 90 percent of the total in andesite and 75 percent in basalt; and that andesites and basalts contain relatively somewhat more carbon monoxide than obsidians and granite but that carbon monoxide is never more abundant than carbon dioxide which is the most abundant gas next to water.

4.2 Theory of Effects of Gases

The presence of spring gases in the form of bubbles, either steam, noncondensible gases, or a mixture of the two is essential to the generation of a geyser eruption. During the early part of the eruptive cycle, the bubbles are very small and well dispersed in the spring water, moving upward together with it at the same velocity. Later the bubbles grow larger and attain upward velocities higher than that of the water. They affect the hydrostatic pressure as if the density of the water had been reduced inversely to the total volume occupied by the water and the bubbles. Thus, the presence of these gases influences geyser action as if the temperature of the water were hotter. Waters highly charged with gases erupt even when water temperatures are well below their normal boiling point.

Under the higher hydrostatic pressure existing at depth, gases such as carbon dioxide, nitrogen, methane, and argon are readily soluble in water and will be carried a certain distance up the geyser tube dissolved in the water (Fig. 4-1). As the waters rise, they will eventually reach some critical depth where the pressure will have decreased to such a point that the gases come out of solution to form small bubbles of gas throughout the liquid much the same as what happens when a bottle of soda water is opened.

As the water rises up the tube, the numerous tiny gas bubbles become larger as the depth decreases and frequently coalesce into larger bubbles. The depth at which bubbles begin to form depends upon the composition and concentration of

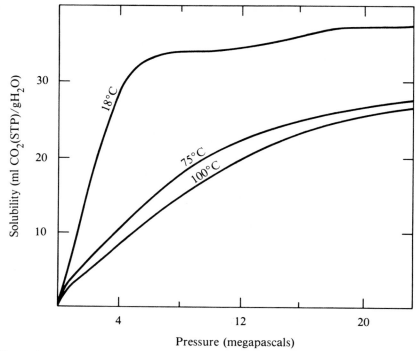

Figure 4-1. Solubility of carbon dioxide in water at various temperatures as a function of pressure. Numbers along curves correspond to temperatures. (Adapted from Miller, 1952.)

the gas and upon temperature. Large whitish swarms of bubbles are observed moving rapidly upward in the basins of many hot springs, frequently evolving into small groups of larger bubbles as the swarms approach the cooler surface water. The presence of these bubbles which lessens the average density of the fluid affects markedly the hydrostatic pressure exerted on the underlying fluid and thereby regulates the temperature and pressure regimes that can exist.

In order to calculate the hydrostatic pressure at any depth exerted by this mixture of bubbles and water, it is necessary to know how its density varies with depth. The density, ρ_{w+b}, will depend upon the relative volumes of water and bubbles given by

$$\rho_{w+b} = \rho_w V_w/(V_w + V_b)$$

where V_w is the volume of water flowing up the tube at depth per unit of time per unit cross section of area; and V_b, the volume of bubbles. The relatively small mass of the gas is neglected.

Assume for simplicity that the only noncondensible gas is carbon dioxide. Immediately after formation, the small occluded carbon dioxide bubbles become

saturated with water vapor and grow to many times their original size. For example, at a temperature of as much as 2°C below the boiling point, the bulk of the volume will be occupied by water vapor, only 6.9 percent of it being carbon dioxide. Since no bubbles form below the critical depth, no heat is lost except a negligible amount by conduction. The situation is different above the critical depth. At first, the bubbles will collapse as they traverse the cooler water, transferring their stored heat to it. As the water in the tube heats up, the bubbles will approach closer and closer to the surface before collapsing. As they rise they expand, due to the decrease in pressure, extracting heat from the surrounding water and setting up a temperature gradient within the geyser tube. The temperature, a minimum at the surface, reaches a maximum at the critical depth, below which it remains essentially constant.

As more and more bubbles form, an increasing volume of water will be displaced which overflows the tube either into a wide-mouthed basin or into a drainage channel. In either case, the hydrostatic pressure is reduced at the lower levels and evolution of gas, which originally began at the critical depth, will propagate rapidly downward to a new level, permitting geysering to develop even though the temperatures are much below the boiling point of water.

The expansion of a spring gas and water vapor-filled bubble is a key element in reducing the hydrostatic pressure. This expansion as it encounters lower ambient hydrostatic pressure during its ascent up the geyser tube can be treated analytically.

To obtain a functional relationship among depth, vapor pressure, rates of fluid flow, and temperature, assume that the gas bubbles are always fully saturated with an amount of water vapor which can form at the ambient absolute temperature, T, of the spring water. The pressure exerted on the bubbles under these conditions is equal to

$$p_o + P - p$$

where p_o is the ambient atmospheric pressure at the surface opening of the tube, p is the vapor pressure of the water, and P is the hydrostatic pressure at depth, d, the depth in the tube at which the bubble is located. Application of the general gas law,

$$pV/T = \text{constant},$$

leads to the following expression for the volume of the gas bubbles, V_b, flowing through a unit cross section of the tube at depth, d, per unit time:

$$V_b = V_{G_o} \left[p_{atm}/(p_o + P - p) \right] \left[T/273 \right] \qquad (4\text{-}1)$$

where V_{G_o} is the volume, reduced to 0°C and standard atmospheric pressure, p_{atm}, of gas flow from the spring. In most calculations, p_{atm} and p_o can be taken as equal to each other.

The heat passing up the tube will be partly stored in the liquid water and partly in the bubble gas, the relative amount taken up by the gas increasing as the water boils, forming larger and more bubbles. Assuming no heat loss in the channel,

the heat content of the fluid, water plus gas, will be constant at all points along the tube and

$$V_w T + p_{atm} V_{G_o} \rho_v \sigma T/273 \, (p_o + P - p_G)$$
$$= V_w T_1 + p_{atm} V_{G_o} \rho_v \sigma T_1/273 \, (p_o + P_1 - p_{G1}) \tag{4-2}$$

where ρ_v is the density of the water vapor in the bubble, σ is the latent heat of evaporation, and V_w is the volume of the liquid water entering the spring per second. The subscript, one, is used to designate respective parameters at a depth of d_1. At great depths, the second term in the right hand side of this equation approaches zero, making

$$p_{atm} V_{G_o} \rho_v \sigma T/273 \, (p_o + P - p) = V_w(T_1 - T) \tag{4-3a}$$

or

$$P + p_o - p = p_{atm} V_{G_o} \rho_v \sigma T/273 \, V_w \, (T_1 - T) \tag{4-3b}$$

which in terms of temperatures t and t_w measured in °C becomes

$$P = p - p_o + p\sigma\rho_{s_o}/(V_w/V_{G_o}) \, (t_w - t) \tag{4-4}$$

where ρ_{s_o} is the density of water vapor under normal pressure at °C, and t_w is the temperature of entering water.

The expansion of a bubble during its ascent can now be calculated using these results. The ratio, α, between the volume of the bubbles and that of spring water, substituting from Eq. (4-1), is given by

$$(V_b/V_w) = \alpha = (V_{G_o}/V_w \left[p_{atm} T/273 \, (p_o + P - p)\right]. \tag{4-5}$$

Substituting from Eq. (4-5) yields

$$\alpha = \left[p_{atm}/273 \, p\sigma\rho_{s_o}\right] (t_w - t). \tag{4-6}$$

The value of the term, $p_{atm}/273 \, p\sigma\rho_{s_o}$, is nearly constant. It is equal to 3.10 at 100°C, 2.66 at 105°C, and 2.28 at 110°C. Thus, α, is primarily dependent on the difference between the temperature at which the spring water enters and the temperature at the point of interest, being numerically equal to two to three times the temperature difference. When $t_w = 110°C$ and $t = 100°C$, the volume of the bubbles, which is 31 times that of the liquid, is sufficient to fill three-quarters of the cross section of the channel and hence disturb the value of the hydrostatic pressure on which Eq. (4-4) is based, even if the velocity of rise is 10 times that of the water. The bubble velocity would have to be 30 times that of the liquid in order not to disturb the hydrostatic pressure which, from practical fluid dynamic considerations, means that the velocity of the liquid could not exceed about 3 cm/s.

At depth of 2 to 3 m where the temperature could reach 105°C without boiling, α would be equal to 15.5. At a bubble speed 10 times that of the water, the bubbles would fill four-sevenths of the cross section of the channel, markedly disturbing the hydrostatic pressure. For higher values, $t_w = 120°C$ and $t =$

105°C, α becomes 40 and the bubbles fill four-fifths of the channel under the same velocity conditions, indicating that a high initial spring water temperature is much more prone to precipitate an eruption and to enhance its violence.

Measurements in Uxahver Geyser in northern Iceland indicate that the temperature of the water when the steam bubbles first start to form below the bottom of the main reservoir is 108°C. Just above this level, essentially at the bottom of the geyser tube at 6 m depth, the temperature is 105.7°C; near the surface, at depth 0.8 m, 100°C. Substituting these values into Eq. (4-6), it is found that the ratio of bubble volume to water volume is 6 near the bottom and 25 near the top, approximately a four fold increase. In Ystihver, a nearby geyser, the volume of the spring gases at 1 m depth and temperature 100°C is 22 times that of the spring water. Water enters Strokkur's main reservoir at a temperature of 115°C, making the volume of the gases near the surface 45 times that of the water.

The spring gases from wells and fumaroles in California contain 1.36 to 3.35 percent by weight of noncondensible gases. Assuming that on the average, geysers contain the same amount, the implication is that 1 g of geyser water at 120°C contains 3/800 g or 3 ml of noncondensible gases in every milliliter of water, making the ratio, volume of water to volume of gas in Eq. (4-4) equal to 0.3.

Geysers in which dissolved and occluded gases play a large role in the eruption process occasionally develop when wells are drilled in regions where there exists a high concentration of subterranean gases. Borehole No. 5, drilled in Iceland in connection with the development of geothermal resources, was such a well. Eruptions 8 m high began to occur as soon as the well had been drilled to a depth of only 15 m even though water temperature at that depth was far below boiling, about 60°C. As drilling proceeded, one or two eruptions would occur per day. Drilling was stopped at 246.5 m, the highest temperature, 98°C, being reached between 223 and 232 m.

Total water discharged from the 9.2 cm diameter borehole during an eruption was about 3000 liters ejected with a velocity of approximately 14 m/s. The eruption lasted 5 min. The discharge rate was therefore 1×10^4 ml/s. During the eruption, the upper part of the borehole would be filled with a mixture of gas and water. The ratio of the volume of water to that of gas can be calculated by relating the volume of water thrown out in a given interval to the volume of the borehole occupied by the water-gas mixture during the same time. If there were no gas mixed with water, a borehole of length, l, and diameter 9.2 cm could hold one second's supply of water where

$$l = 1 \times 10^4/4.6^2 = 150 \text{ cm}.$$

The length of pipe emptied per second of fluid, gas and water, was 1400 cm, making the ratio of the volume of the gases in the topmost part of the borehole during an eruption approximately 10 times that of the water.

The effects of the pressure of bubbles on temperature and pressure distributions are fairly complicated but can be treated analytically for the idealized case of a single vertical channel having uniform cross section. The equations relating

to geyser action derived in Sec. 3.4 can be extended to take into account the influence of noncondensible gases. Assume that the volume of gas bubbles passing upward by any point, A, through a unit cross section, is V_b. At the same time, V_w of $T°$ water is passing the same point. For simplicity, assume only one type of noncondensible gas is present, part of which is dissolved in the water and part resides in the bubbles. The whole volume of noncondensible gas passing point A at depth, d, per second is the sum of these two parts. Reduced to standard conditions, 0°C, and sea level atmospheric pressure, the total gas, V_{G_o}, is constant and given by

$$V_{G_o} = [273(p_o + P - p)/p_{atm}] [V_b + KV_w] = \text{constant} \qquad (4\text{-}7)$$

assuming no accumulation of gas along the channel. Henry's constant, K, is the amount of gas in solution in water at hydrostatic pressure. The rate of flow of water plus steam, Q_{w+s}, per unit area of cross section is given by

$$Q_{w+s} = \rho_w V_w + 273\, V_b\, p\, \rho_{s_o}/p_{atm}\, T. \qquad (4\text{-}8)$$

Thus, solving for V_w gives

$$V_w = Q_{w+s}/\rho_w - (273\, V_s\, p/p_{atm}\, T)\, (\rho_{s_o}/\rho_w). \qquad (4\text{-}9)$$

Assuming that the rate of transport of heat up the channel by the hot fluids is constant and disregarding any heat losses, usually justifiable, then from Eqs. (4-3b) and (4-9)

$$V_b = p_{atm}\, T\, Q_{w+s}\, (T_1 - T)/273\, p\sigma\rho_{s_o}. \qquad (4\text{-}10)$$

Substituting from Eqs. (4-7) and (4-9) in Eq. (4-10) and reducing, gives for the hydrostatic pressure

$$P = p - p_o + (V_{G_o}/Q_{w+s}) \left[273(\sigma + T_1 - T)K/p_{atm}\sigma\rho T + (T_1-T)/p\sigma\rho_{s_o}\right]^{-1}. \qquad (4\text{-}11)$$

Taking the density of water as one and noting that $\sigma = 540$ cal/g, the quantity $(\sigma + T_1 - T)/\sigma$ becomes $\left[1 + (T_1 - T)/540\right]$ which is approximately equal to one since $(T_1 - T)$ is seldom more than 10 to 20 degrees, then

$$P \approx p - p_o + (V_{G_o}/Q_{w+s}) \left[273\, K/p_{atm}\, T + (T_1 - T)/p\sigma\rho_{s_o}\right]^{-1}. \qquad (4\text{-}12)$$

The term containing K is usually small so that

$$P \approx p - p_o + (V_{G_o}/Q_{w+s}) \left[p\sigma\rho_{s_o}/(T_1 - T)\right]. \qquad (4\text{-}13)$$

This equation establishes an equilibrium condition. Steam bubbles will just begin to form and remain stable when the following equality is reached:

$$p = (P + p_o)/(1 + \theta) \qquad (4\text{-}14)$$

where, for convenience,

$$\theta = (V_{G_o}/Q_{w+s}) \left[\rho_{s_o}\, \sigma/(T_1 - T)\right], \qquad (4\text{-}15)$$

the coefficient of p in the last term of Eq. (4-13). Except for the case where no noncondensible gas is present and θ is zero, θ is always positive, increasing with an increase in the amount of gas present. Inspection of Eqs. (4-14) and (4-15) indicates that the higher the gas content the lower the vapor pressure, and hence temperature, needed for bubbles to begin forming, assuming that the same hydrostatic pressure obtains. Contrariwise, for water of a fixed temperature, hence constant vapor pressure, the value of hydrostatic pressure at which bubbles will form is higher, implying that eruptions will begin at lower depths in gassy geysers than in nongassy ones.

The depth, d_g, at which ebullition would begin and proceed to depth, d_e, can now be determined in the same general manner as was done in Sec. 3.4 where the presence of noncondensible gas was not taken into account. The increment of hydrostatic pressure exerted by a gassy water column will be given by Eqs. (3-5), (3-6), and (3-7), which, after suitable integration and simplification become

$$d_e = p_{T_1} - p_o + (p_{atm}T_1 V_{G_o}/273 \text{ K } Q_{w+s}) + \Delta d_g$$
$$= d_g + \Delta d_g \qquad (4-16)$$

where d_g is the depth at which the gas occluded in the spring water is able to release itself and form bubbles if no gas evolution has disturbed the hydrostatic pressure of the overlying waters, and Δd_g is the distance the release works itself down before stopping at d_e. The difference between d_g and d_1, when no gas is present, is

$$d_g - d_1 = (p_{atm}T_1/273 \text{ K}) (V_{G_o}/Q_{w+s}) \qquad (4-17)$$

indicating that the difference between d_g, which is the greater of the two, and d_1, is directly proportional to the amount of gas present. Similarly, Δd_g is greater than Δd_1.

In summary, the presence of spring gases on geyser action has substantially the same effect as the presence of hotter water. Large amounts of gas greatly provoke geyser action to the extent that springs whose water temperatures are well below the boiling point behave as geysers. Indeed, numerous drilled wells erupt regularly even though their waters are as cool as 15°C.

4.3 Gassy Geysers

A most spectacular gassy geyser developed in 1878 in an oil and gas region of Pennsylvania (Fig. 4-2). Because of the high hydrostatic pressure existing in the region, the oil was usually projected as a jet from a freshly drilled well. One such well near Kane, in McKean County, was drilled to a depth of 600 m. It was abandoned because of the small yield of oil but later became a gas-driven water geyser. It projected a column of gas-saturated water at intervals of from 10 to 15 min to heights ranging from 30 to 45 m. The gas being an inflammable hydrocarbon was often lit at night and "the antagonistic elements of fire and water were

Figure 4-2. Facsimile of old etching of water-and-illuminating-gas geyser in Pennsylvania. (From Dana, 1894.)

promiscuously blended, at one moment the flame being almost extinguished but only to burst forth the next instant with increased energy and greater brilliancy.'' The gas probably came from a vein heavily saturated with gas that had been struck at the 430 m depth during drilling.

Several other wells in the eastern USA oil-producing regions have become

similar geysers. As early as 1833, a salt well in the Ohio valley threw jets of water and gas at intervals of 10 to 12 hr, to heights from 15 to 30 m.

Somewhat similar geysers developed in the Boja-Dag region of the USSR on the Pribalk Laskaja plain while putting down test holes for oil and gas. Copious quantities of water were found in some. The emission of hydrocarbon gases is characteristic of the Boja-Dag springs and the well water is quite warm, 40°C to 60°C.

The water flowing from one geysering well was at a temperature of 69°C. It mounded up 30 to 40 cm in the center of a small pool, 5 m in diameter and about 1.5 m deep. The steady flow of water, 10 liters/s, was accompanied by an intensive efflux of hydrocarbon gases, making it appear to be boiling vigorously. About every 13 hr, the flow of water changed abruptly, a hollow subterranean rumble was heard, and a column of water was thrown 12 m in the air. After playing for 8 to 9 min, the discharge of water stopped abruptly and the pool and well completely dried up. About 50 min later, the water would reappear in the well at a depth of 18 m and in another 15 min, accompanied by much splashing, the water would rise to the top of the well, filling the pool and forming its characteristic churning mound of water. Over a period of time, the interval between eruptions increased, being originally 45 min in 1953, increasing to 6 hr in 1956, and to 13 hr in 1959. Its present activity has not been reported. The decrease in activity was probably due to depletion of the gas reserves in the region.

The Crater Hills Geyser, in the eastern part of Yellowstone, was erupting at temperatures far below boiling. Its behavior, as described a few years ago, was markedly different from that of a steam-activated, hot water geyser. The vent opens into a fairly deep, 5 by 7 m pool. The eruption cycle followed a quite regular pattern. Between eruptions, the water sat quietly in the basin, filling it to within 0.5 m of its top. The eruption began suddenly without preplay. Violent boiling started at the center of the basin creating a long series of rapidly occurring 1 to 2 m high splashes. This continued for about 30 min, with all of the water falling back in the basin. During the eruption, the water in the basin gradually rose. When it reached the rim, the geyser stopped erupting and the water became quiet. It remained so for about 2 min, with the water gradually receding to the 0.5 m level, preparing for the next eruption which occurred at this point. Since the temperature of the water was only 73°C, the geyser must be powered by spring gases other than steam. The waters are strongly acid, pH 3.9, contrasting with the alkaline waters of most hot water geysers. The other contrast is that the eruption was much longer than the interval.

Crystal, a gassy cool water geyser, is activated primarily by the evolution of carbon dioxide. The water in the pipe just before eruption is about 15°C. The geyser maintains a fairly constant pattern, erupting in the late 1960s on a 5 hr schedule, playing a massive and impressive stream of water for 5 to 10 min to a height of 60 m (Fig. 4-3). The geyser is located some 8 km downstream from Green River, Utah, on the eastern bank of Green River. The jet issues from a 45 cm diameter casing emplaced to a depth of a few meters. Presumably this geyser

Figure 4-3. Crystal Geyser, Green River, Utah, in eruption. Pipe from which jet issues is 45 cm in diameter; jet is 50 m high.

became active when a geophysical exploration group put down a fairly shallow hole looking for geothermal energy, having been attracted to this spot by a geyser-like mound of mineral deposits. Most of the natural vents have been sealed by internal accretion of these. A second well is located a few meters away. It erupts in sympathy but only plays to a height of 1 to 2 m.

Immediately following the main eruption, the water in both vents is about 8 m below the flanges. A small second burst, which occurs about 25 min later, comes with very little warning, as, with a rushing sound, the water in the tube rises rapidly from its 8 m depth to the top of the vent pipe in about 2 s and then throws a jet about 2 m high. After this, the water recedes down the wells, completely out of sight. The smaller well begins to overflow 80 to 90 min after the secondary eruption; about 3 hr after the eruption, the main vent overflows for the first time with an increase in foaming and hissing in the pipe. This and subsequent over-

flows, which come at hourly intervals and last for a minute or so, increase in vigor. The one just ahead of the precipitating overflow lasts about 2 min. The eruptive episode takes a while to get going, the gas-laden water sometimes rising and falling two or three times. Once fully underway, the eruption reaches its maximum height in only 4 to 5 s. Since it has no steam associated with it, the water jet appears as a gradually tapering, very symmetrical obelisk. The jet begins gradually to drop, retreating to the top of the vent where it splashes a few seconds and is then sucked down the well out of sight. Several seconds later, the water calmly rises to the 8 m level in the main well. A few minutes later, the air in the vent suddenly starts to whistle and white foaming water is briefly thrown several meters upward from its opening.

The much smaller carbon dioxide-activated Roadside Geyser is located in a flat alluvial basin about 48 km northwest of Green River. It came into being about the turn of the century when a 13 cm diameter well was drilled to a depth of 1000 m. The well began erupting and as late as the mid 1970s was throwing a very narrow jet of water at intervals of 40 min.

Both of these geysering wells are close to regions where subterranean carbon dioxide and helium exist in high enough concentrations to be mined profitably. Thus the water of the wells in this area could be expected to be saturated with dissolved carbon dioxide under the high ambient pressures existing at depth where its solubility in water is high (Fig. 4-1). Reduction in pressure as the water rises up into the well would lead to release of carbon dioxide, commencing a chain of events similar to the action of gassy, hot water geysers.

There are several cold water geysering drilled wells in Czechslovakia. The best known of these is at Herlány in the eastern part of the country. While Bad Herlány has been known since the seventeenth century, the 404 m deep borehole from which the geyser erupts was drilled and cased around the 1880s. The casing in the hole, extending down 351 m, reduces from 40 cm in diameter to 13.

Regular eruptions formerly occurred about every 9 hr. The interval in 1973 had lengthened to about 22 hr. An eruption is usually heralded by a peculiar deep thumping sound. This is followed by an efflux of gas from the orifice developing into rising white foam. Suddenly a thin water jet shoots up to a height of 18 or 20 m and continues to play for approximately 30 min.

Although the temperature of the ejected fluid is only 20°C, carbon dioxide gas constitutes a far larger volume of it than the liquid water. The chemical composition of the water is close to that of an alkaline hot spring.

A second well was developed in Slovakia at Perši, close to Lučenec. It is smaller, has a shorter interval, and contains no appreciable amount of carbon dioxide. Another at Neuenahr erupts carbon dioxide-rich 47°C warm water from a depth of 90 m for 1.5 to 2 hr, followed by a quiet period of about equal length.

There are also some cold water geysers in France of which the best known is that of "Martes d'Artieres."

A few investigators have attempted to attribute the action of these geysers to siphoning of water-filled underground cavities topographically higher than the

geyser's orifice. And, indeed, model geysers which operate in this way have been constructed. It seems more likely that the correct explanation lies in the discharge of dissolved gases, invoking in general the process discussed in Sec. 4.2.

CHAPTER 5

Chemistry of Geothermal Waters

5.1 Water Sources

At least 95 percent of hot spring waters is meteoric, precipitation from rain and snow, and originally free of mineral content. Any remainder could be magmatic. The substantial quantities of chemicals they carry when they reappear at the surface are picked up during their circuitous subterranean passages. The extreme diversity in the composition of hot spring waters results principally from chemical reactions with the particular rock minerals through which they pass, with dissolved gases, and with atmospheric gases present. Further, mixing and diluting with cold, shallow, less mineralized meteoric water or those with different dissolved components as well as the boiling off of water affect their composition. Geochemical analyses lead to somewhat ambiguous conclusions although the bulk of the evidence points to rather extensive mixing in most cases.

Analysis of Steamboat Springs waters suggest that saline, magmatic water has been diluted to varying degrees by meteoric water of low saline concentrations. Heavy precipitation there has a marked effect on the discharge rate, salinity, and temperature of the waters. Springs close together in the same basin may differ greatly from one another and the same spring often varies from time to time apparently caused by differences in the amount and type of percolating near-surface water that reaches it.

Some geyser reservoirs are probably fed by rivers, lakes, or even the ocean as at Atami in Japan where the spring waters are highly saline. Often there are several sources of geyser water, the most abundant being relatively cool shallow ground water with smaller amounts of much hotter water returning to the surface after it has percolated to great depths and become heated. But a geyser having only a single source of hot water can perform perfectly well. Model geysers constructed in this way have all of the performance characteristics of natural geysers. On the other hand, small Minute Geyser in Norris Basin, whose plumb-

ing and heating system is completely exposed, has as a source of heat steam injected through vents in the floor of the shallow geyser reservoir which is already filled with ground water.

Magmatic or juvenile water is water that is given off by the magma as it cools. The proportion of it in geyser waters, as seen by analyzing the relative abundance of the two hydrogen isotopes in them, is very small, not more than 5 percent and probably much less. The much more abundant isotope, H, has an atomic mass of one; the other, D (deuterium), has an atomic mass of two. In average sea water, the ratio between the relative amounts, H/D, of the two is 6400. Most of the deuterium resides in water molecules having the composition HDO. The scarcity of deuterium insures that few D_2O molecules will form. The HDO, being heavier than the H_2O molecules, has a lower vapor pressure and a higher proportion of H_2O molecules will evaporate from the ocean. Thus the resulting rain, when it falls, will contain less deuterium than in sea water. Analysis shows the H/D ratio in rain water, river water, and tap water is 6800; essentially the same value of the ratio is found in geothermal waters, further support for the hypothesis that the bulk of geothermal water comes from precipitation. The inference is certainly reasonable. Rain and snowfall in Yellowstone, Iceland, Kamchatka, and New Zealand are fully ample to account for the hundreds of springs that exist.

The number, size, and depth of springs in a given area usually relate directly to the extent and slope of the catchment area, another indication that the bulk of water reaching the springs is percolating ground water. Groups of high discharge springs, generally the larger and deeper ones, are located in well watered basins whereas those located on hillsides and in small basins and ravines are smaller. Where there is very little ground water superheated jets, fumaroles, are likely to develop.

In many basins the variety of types of springs indicate a highly localized and uneven distribution of water. A deep spring will develop in otherwise quite dry ground and mud pots in wet ground. The explanation for this lies in variations in the permeability of the ground. Sand and gravel tend to distribute the water uniformly whereas solid rock restricts the circulation to joints, seams and fractures.

5.2 Composition of Geothermal Fluids

Although some of the mineral matter in hot spring waters is derived directly from magma, most of it comes from dissolution of the primarily volcanic rocks through which the waters pass. Boiling springs and geysers usually discharge chloride water which is neutral at depth but becomes alkaline by the time it reaches the surface.

The compositions of hot spring waters are strongly influenced by the type of hot ground in which they are situated and consequently range widely. They may be classified as follows:

• Alkaline waters, containing several hundred ppm of bicarbonates and

chlorides, and much lesser amounts of boric acid and fluorine. Sometimes carbonates are present. The sulfate content is low. Generally the silica content ranges from 200 to 300 ppm, except at Norris Basin where it is quite high, 700 ppm. Sodium, potassium, and lithium always seem to be present, even if only in small amounts.

• Either acid or close to neutrality, predominately sulfate waters with chlorides and fluorides present only in very small amounts. The concentration of sulfuric acid, usually about the only acid present, ranges from less than 10 ppm in most of the acid waters to 80 ppm in a few cases, i.e., from less than 0.5 g/liter to 4 g/liter. The silica content is characteristically from 200 to 300 ppm. Ammonia is often present but in extremely variable amounts. Sodium, potassium, calcium, and magnesium are the principal metals present, with sodium by far the most abundant. Only traces of iron and aluminum are present.

• Waters characterized by much calcium bicarbonate and very little silica. This type is only rarely found in Yellowstone.

The great majority of geysers in Yellowstone are clear, deep, sinter-lined alkaline springs. A few play from fissures in the rhyolite rock with little accumulation of sinter. The waters are as limpid as a mountain stream and flow relatively fast. The brilliant blue color of the hot pools and geysers, the unusual clarity of the water, appreciable basin depth, and light-colored basin walls and bottoms are characteristic of the most beautiful of the springs. Reflection of the sky cannot account for the blue color since it is almost as bright on a cloudy day. Much of it is attributable to preferential absorption of light according to wave length. The effect of an absorption band beginning in the near infrared at 6000 Å can be observed even through a one meter depth of pure water. Other factors probably contributing to the effect are selective scattering and diffusion of light by fine particles suspended in the water, the depth of the spring's basin, the reflecting character of its bottom, and the position of the observer. Usually only alkaline springs exhibit the effect. The few acid springs are turbid and muddy due to suspension of dark mineral particles that remain in the spring as a result of low rate of water flow.

Occasionally the blue of the water is modified by another substance. For example, the brilliant green of Emerald Pool, Upper Geyser Basin, is a combination of the blue water and the yellow algae on the walls. Suspended free yellow sulfur occasionally imparts a less pleasant green to the water. The jet black-ink appearance of the waters of several springs in the eastern areas of Yellowstone is caused by suspended iron sulfide particles.

The more heavily mineralized thermal waters at Steamboat Springs contain 2000 to 8000 ppm of mineral matter, mainly sodium chloride, carbonates, carbonic acid, silicon dioxide, sulfate, and boron. They range from slightly acid to moderately alkaline.

The pH of the Japanese geysers, listed in Table 5-1, range from highly acid, 2.4, to alkaline, 8.7, biased somewhat toward the alkaline side. This is the same range observed for the nonerupting hot springs indicative that the pH of the

Table 5-1. pH and Mineral Content of Japanese Geysers and Geysering Wells (Adapted from Iwasaki, 1962)

Geyser	pH	Solid Residue after Evaporation (g/liter)
Hokkaido		
Noboribetu Geyser	6.5− 7.1	7.7− 9.1
Sikake Geyser	8.0	3.8− 3.9
Aomori Prefecture		
Osoreyama	5.8− 6.3	11.0− 14.4
Syuraozigoku Geyser	6.4	
Miyagi Prefecture		
Onikobe		
Megama Geyser	2.4	
Fukiage Geyser	8.2− 8.5	
Takagame Geyser	~8.4	
Miyazawa Geyser	6.8− 8.2	
Benten Geyser No. 1	8.0− 8.3	
Benten Geyser No. 2	8.4	
Benten Geyser No. 4	8.1− 8.7	0.1 (No. 5 Geyser)
Narugo		
Tosenro Geyser (large)	7.7	1.9− 2.0
Tosenro Geyser (small)	7.7− 8.0	
Kanetyu	7.7	1.9− 2.3
Yamagata Prefecture		
Mogami Geyser	8.0	
Fukusima Prefecture		
Tutiyu Geyser	7.1− 8.5	1.3− 1.4
Nigata Prefecture		
Matunoyama Geyser	7.4− 7.7	15.5
Hyogo Prefecture		
Arima Tenjin-no-yu Geyser	6.3− 6.6	73.0
Oita Prefecture		
Beppu		
Tatumakizigoku Geyser	2.5	3.2− 3.5
Yufuin		
Yosimoto Geyser	7.2	
Kimoto Geyser	7.5	1.1− 1.2
Kumamoto Prefecture		
Yunotani (Volcano Aso)		
Aso Geyser	5.0	0.7−0.8
Hōkōzigoku Geyser	5.4	0.6−1.6
Common Hot Springs		
Range		0.1−85.9
Average		3.9

waters must not be important in geyser action, a conclusion supported by other observations. Geysers in Yellowstone and New Zealand issue from rhyolite; in Steamboat Springs from granite; in Beowawe, basalt; in Iceland, both rhyolite and basalt; and in Japan, andesite and andesitic rocks. The distribution of mineral matter in the geysers also corresponds to that of collocated hot springs. The table also lists the amount of solid residual matter after evaporation for several individual Japanese geysers and a summary of similar data for hot springs.

The compositions of the mineral matter typically dissolved in waters taken from geyser areas through the world are listed in Table 5-2 and for the Great

Table 5-2. Chemical Compositions of Geyser Waters (ppm) (Adapted from Iwasaki, 1962; Rowe et al., 1973)

	Steamboat Springs, USA	Beowawe, USA	Tatumakizigoku, Japan	Papakura, New Zealand	Great Fountain, Lower Geyser Basin, YNP, USA	Sapphire Spring, Upper Geyser Basin, YNP, USA	Little Whirligig, Norris Basin, YNP, USA	Clepsedra, Lower Geyser Basin, YNP, USA
NH_4	4					0.2		3
Na	744	282	829	395	334	450	349	380
K	77		132	53	14	16	83	13
Ca	6	tr	78	1		0.3	3	0.5
Mg	tr	0	28	0		0.01	1	0.01
Al	0	tr	1				2	
Fe	tr	tr	12				1	0.05
CO_3	60	tr		39				
HCO_3	248	512				595		480
SO_4	127	91	626	58		17	113	22
S_2O_3	2							
Cl	978	70	1255	548	340	308	607	325
F	2?			9		30	3	22
B_2O_3	179							
SiO_2	343	418	266	203	325	334	420	414
Li				3		2	5	3
Mn			6					
HBO_2				22				
B						3	9	4
H_2S						3	4	1
pH						8.5	7.1	9
Re	2770	1307	3260					

Geysir in Table 5-3. The mineral matter in the Japanese geyser waters is much larger than that in those elsewhere. At Yellowstone, it is relatively small, ranging from 1.3 to 1.6 g/liter; it is about the same, 2 g/liter, in New Zealand and Iceland whereas in Japan it is sometimes greater than 10 g/liter.

In certain geysers substantial and very repeatable variations in chemical composition of the water occur as they discharge during eruptions. Contrariwise, any variation in the chemical composition of the waters ejected by perpetual spouters is very small. The variations in composition are due partly to mixing of waters from various sources during an eruption and partly to the increase in concentration of mineral matter in the residual water after flashing. The effect of flashing during an eruption of Pohutu Geyser is evident in Table 5-4, which lists the composition of the waters at three stages during an eruption. The chloride varies by about 5 percent. The variations in chloride content in some of the Japanese geysers (Table 5-5) is much larger, almost a factor of two at the Miyazawa Geyser.

Table 5-3. Chemical Composition of the Waters of the Great Geysir (Adapted from Bunsen, 1848)

Chemical	(ppm)
SiO_2	510
$NaCO_3$	194
NH_4CO_3	8
Na_2SO_4	107
K_2SO_4	48
Mg_2SO_4	4
$NaCl$	252
Na_2S	8
H_2CO_3	56

Table 5-4. Analysis of Pohutu Geyser Water During an Eruption (Adapted from Lloyd, 1975)

Sample	Cl^- (ppm)	F^- (ppm)	NH_4 (ppm)	HBO_2 (ppm)	Cl/F (ppm)	Cl/ABO_2 (ppm)	$Cl/(NH_4)$ (ppm)
First water ejected	582	8.3	0.38	21.5	37.8	33.4	777
Half through eruption	604	8.5	0.76	22.7	38.0	32.9	404
End of eruption	616	8.7	1.87	22.9	38.2	33.2	167

Table 5-5. Variations in Chloride Content in Japanese Geysers During an Eruption (Adapted from Iwasaki, 1962)

	Variation in Chloride Content (percent)
Geysers	
Miyazawa	92
Narugo Tosenro (small)	27
Narugo Kanetyu	15
Tutiyu	8
Narugo Tosenro (large)	7
Sikake	3
Matunoyama	0.5
Continuous Spouters	
Serrami	1.4
Mine	1.0
Tamagawa	0.6

5.3 Water Movements and Contacts: Geothermometry

A typical geyser area is a basin several kilometers in extent, usually a mass of loose gravels resting on heavily jointed and fractured rock. All this in turn lies over a plug of hot magmatic rock. In the Upper Geyser Basin, drilling has indicated that the top layer, siliceous sinter, is about 6 m thick; below it there is about 70 m of glacially emplaced rhyolite gravels underlain by fractured rhyolite (Fig. 5-1).

These basins are excellent collectors of meteoric water, a portion of which percolates downward, often hundreds of meters, through the sediments and rocks. There it is heated, becomes lighter, and driven by its buoyancy, returns to the surface to heat geyser reservoirs and hot pools and to escape as steam from fumaroles. The remainder stays close to the surface and fills the geyser reservoirs and hot springs after having been only somewhat warmed.

Temperature-depth profiles taken down the 70 m of surficial rhyolite gravels in Upper Geyser Basin suggest that here the water moves in more or less horizontal currents, probably part of large convection cells. Below the 70 m level, they presumably are constrained to follow the existing fractures and joints which trend heavily toward a vertical direction.

The speed of movement of the deeply circulating water can be expected to be extremely slow. It can take at least a few decades and perhaps longer for the water to complete its traverse. Thus short term variations in the amount of precipitation will have little effect on its movement. The speed will vary greatly from point to point, depending upon the permeability of the sediments and rocks, and the temperature, density, and viscosity of the water.

It has been shown experimentally that the rate of flow of water, q, through a

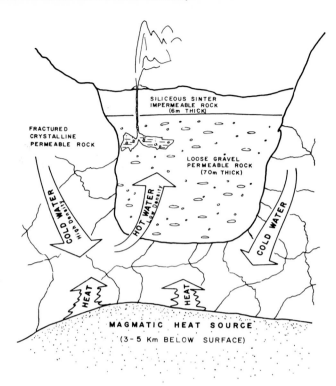

Figure 5.1. Geologic cross section of typical geyser basin, especially applicable to the Yellowstone basin. (Adapted from White, 1967.)

unit cross section of soil, gravel, and porous and fractured rock, is proportional to the difference in pressure existing between the two faces of the section. The coefficient of proportionality, K, is defined as the permeability. Analytically, the flow is given by

$$q = K\Delta p$$

where Δp is the gradient of pressure across the section through which flow is taking place. Thus permeability is a measure of the ease with which water will flow through soil and rock masses and must be taken into account when considering water flow through them. It depends upon such factors as grain size, porosity, and size, number, and spacing of the joints and fractures. It also depends upon the viscosity of the fluid which in the case of water is some 20 fold less at the high temperatures found in geothermal areas than at normal ambient temperatures. Water tends to follow joints, fractures, faults, and open solution channels, and as the character of these change, numerous individual water flows will develop below the gravel.

Geochemical and geophysical evidence indicate that at the Upper and Lower Geyser Basins, at least three discrete hydrothermal systems exist near the surface

and these can account for the great diversity of the thermal features. Each is at a different temperature and contains fluids of different compositions. At greater depth, 2 to 4 km, these discrete systems are interconnected to form a single aquifer filled with 340°C to 370°C brine containing approximately 1000 ppm of sodium chloride. This aquifer feeds hot water and steam upward along appropriate channels into the near-surface systems. As it flows upward from the great depth, it undergoes a substantial change in its mineral content: it reacts with rocks, loses steam, and becomes diluted when it mixes with the cooler overlying meteoric waters stored in the discrete hydrothermal systems. The evidence for the existence of the large aquifer and its associated smaller ones is derived from the chemical and isotopic compositions of the waters, from use of isotopic geochemistry, from relationships between the chloride concentrations and subsurface heat contents of thermal waters, and from geophysical investigations.

Geothermometry, the measurement of temperature deep within a geothermal area, has depended primarily on the fact that mineral solubility and concentration equilibrium are temperature dependent. Two chemical geothermometers and two isotope-abundance geothermometers have been used with considerable success. One chemical one is based on the variability of the solubility of silica, SiO_2, which varies with temperatures as shown in Fig. 5-2; the other, on the dependence of the ratio of sodium to potassium and calcium, Na/K and Ca, in the dissolved minerals. The two isotope geothermometers rely, respectively, on the relative abundance of carbon and oxygen isotopes.

Temperatures determined by the silica analysis are somewhat unreliable since the solubility of SiO_2 is different for its various polymorphs; the results of analyses are strongly influenced by the methods and conditions of sample collection and storage, and by the treatment procedures before analysis. A correction has to be made for the steam that forms because of reduction in pressure on the fluid as it moves upward. Further it is generally assumed that the equilibrium

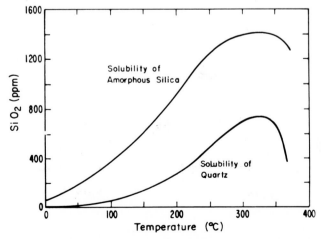

Figure 5-2. Solubility of fuzed silica and crystalline quartz in water as a function of temperature. (Adapted from Truesdell and Fournier, 1975.)

situation that exists at depth persists until the fluid reaches the surface to be sampled. Obviously this assumption is not strictly valid so that only minimum temperatures are obtained.

For the usual, more complex situation where hot and cold waters mix in the geyser reservoir, graphical procedures based on heat and silica balances are generally used to estimate the temperature and proportion of the hot water component to the cold water. It is assumed either that (1) no heat or steam is lost from the hot water component before mixing, or that (2) steam separates from the hot water component at an intermediate temperature before mixing. Either way it is assumed that no heat is lost after mixing, that the initial silica content of the deep hot water is representative of the solubility of quartz, and that the dissolved silica content remains constant during mixing. Specifically the procedure used by Truesdell and Fournier (1977) is as follows. If there is not loss of steam before mixing (assumption 1), then:

● determine or estimate the temperature and silica content of nonthermal ground water in the area and after relabeling °C as calories, plot point A (Fig. 5-3), which is the silica solubility versus temperature curve replotted as solubility versus enthalpy;

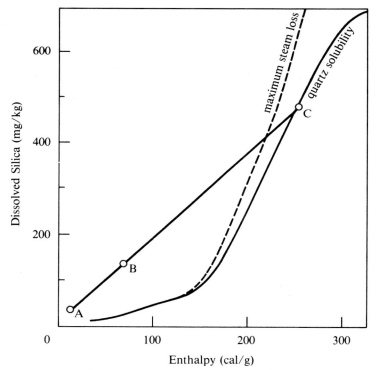

Figure 5-3. Solubility of quartz in water as a function of enthalpy. Graph and construction to be used when assumption is made that no steam or heat has been lost before mixing. (After Truesdell and Fournier, 1977.)

• plot silica content and enthalpy of the warm spring water on the same graph as point B;

• connect the two points, A and B, with a straight line, extending it to intersect the quartz solubility curve, point C, which corresponds to the silica content and enthalpy of the deep hot water component;

• determine the temperature of the hot water component from its enthalpy by using the steam data on Fig. 5-4; and finally

• determine the fraction of hot water in the warm spring by dividing the distance AB by AC.

If there is a loss of steam from the hot water before mixing (assumption 2), then:

• follow the same procedure as above, plotting the temperature and silica contents of the warm and cold waters as points A and D (Fig. 5-5);

• extend the line along which these lie until the enthalpy corresponds to that of the liquid water mixture from which the steam escaped, point E; and finally

• draw a horizontal line from point E so as to intersect the maximum steam loss, point F, and then drop vertically to the silica solubility curve to point G.

The enthalpy of the hot water component before and after boiling has occurred is that at point F and the original silica content before any loss of steam that at point G. The fraction of hot water accounting for steam loss in the warm spring is

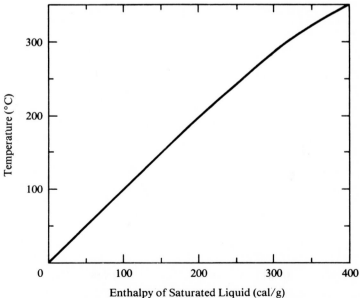

Figure 5-4. Temperature-enthalpy curve for liquid water in equilibrium with steam. (After Truesdell and Fournier, 1977.)

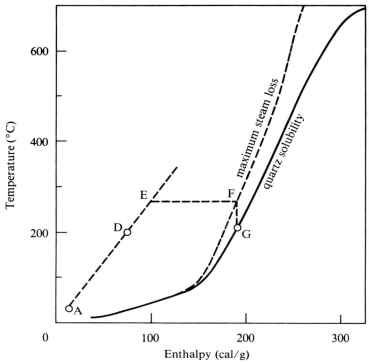

Figure 5-5. Construction to be used with dissolved silica-enthalpy curve when assumption is made that steam separated at 100°C from the hot water component before mixing. (After Truesdell and Fournier, 1977.)

obtained by dividing the distance AD by AE (Fig. 5-5) and the weight fraction of original hot water lost as steam before mixing, x, is given by

$$x = 1 - (\text{silica content at G})/(\text{silica content at F}).$$

The sodium-potassium-calcium ratio method used for determining the temperatures of deep waters is entirely empirical, being based on the relative molar concentrations of sodium, potassium, and calcium in thermal waters over the temperature range from 4°C to 340°C. The data are best fit by the straight line (Fournier and Truesdell, 1974)

$$\log(Na/K) + \beta\log\left[(Ca)^{1/2}/Na\right] = 1647/(273 + t_o) - 2.24$$

where β is $1/3$ for water temperatures above 100°C and $4/3$ for ones below, and t_o is water temperature expressed in °C.

The sodium-potassium ratio method appears to be more reliable than the silica method, giving temperatures believed to be accurate to within 15°C or 20°C.

Average subsurface water temperatures determined from silica geothermometry and Na-K-Ca geothermometry are listed in Table 5-6 for several basin

Table 5-6. Chemical Geothermometer Compared with Calculated Average Temperatures at Yellowstone (°C) (Adapted from Truesdell and Fournier, 1975)

Geyser Basin	Determined by SiO_2 content	Determined by Na/K/Ca ratio	Calculated, assuming mixture of waters
Lower and Midway	179	162	210
Norris	210	251	276
Heart Lake	196	188	221
Upper	195	186	230

areas in Yellowstone. Calculated temperature values are given in the same table assuming that subsurface cooler water, containing only small amounts of mineral matter, mixes with the deeper more saline water.

The average chemically determined temperatures at Yellowstone, except for Norris Basin, appear to be too low. At such temperatures, amorphous silica could be expected to be deposited down within the springs and not carried out, which is not the case. However, the calculated mixture temperatures for the hot components are equal to or in excess of that temperature at which silica would remain in solution. This suggests that the calculated mixing temperatures are the more realistic of the two sets of temperatures, the chemical temperatures being significantly influenced by re-equilibration near the surface. The high acidity and rapid upflow of thermal water at Norris reasonably could inhibit deposition of silica near the surface. Furthermore, even the mixture temperatures are probably somewhat lower than those of the hottest water of the mixtures because of earlier mixing with other cooler waters.

The chloride contents of the waters of some of the springs at Yellowstone have been used to assess the general validity of the chemical methods. As hot water flows upward from depth and boils, the percentage content of the non-volatile chlorides (NaCl, KCl, etc.) in the residual water usually increases. The proportion of water transformed into steam, which can be computed, depends upon the temperature of the water. The silica content behaves in a more complex way; it may act similarly and its concentration increase, or it may precipitate and thus decrease the amount present. At the elevation of Yellowstone where the boiling point is 93°C, 23.9 percent of 217°C water will flash into steam. For example, at one spring, the chloride content at the surface was observed to be 46 mg/liter. The concentration, C_{T_a}, at temperature T_a deep in the aquifer, is given by

$$C_{T_a} = T_a (1 - 0.239) = 355 \text{ or } T_a = 466°C.$$

Subsurface temperatures of numerous Yellowstone springs established using the silica method are entirely consistent with temperatures measured recently in holes drilled to depths of a few hundred meters in corresponding areas.

Analysis of hot spring waters for the relative abundances of isotopes is be-

lieved to be an even more reliable way than Na/K/Ca analyses for determining the source temperatures of the waters. Two istope suites have been used at Yellowstone: (1) determination of the distribution of carbon 13 between CO_2 and CH^4 ($\Delta C^{13}[CO_2, CH_4]$) in the discharging gases and (2) the oxygen isotopic compositions of spring waters and their dissolved sulfates. As a result of isotope exchange with rock minerals, the hot spring water increases its relative concentration of oxygen 18 substantially above that with which it started as meteoric water. Analysis of Yellowstone waters by both methods have yielded temperatures ranging from 244°C to 390°C from the first method and 350°C from the second.

5.4 Solubility of Rocks; Rock Alteration by Thermal Waters

On an average, the hot spring waters at Yellowstone discharge 1.32 g of dissolved mineral water per liter, corresponding to a total of 352,000 kg/day. Analysis of this material certainly shows that the bulk of it is derived from the rocks with which the underground water comes in contact. A typical rhyolite in Yellowstone is about 70 percent silicon dioxide, the remainder consisting mostly of oxides of the light metals. The compositions of such rhyolites are listed in Table 5-7. The slight amounts of boron, usually in the form of boric acid, and arsenic found in the waters of many geothermal areas including Yellowstone are exceptions. They must be derived from the underlying magma itself as rhyolites do not contain even small amounts of either.

The silica in thermal waters is derived from solution of silicon dioxide, sometimes in the form of crystalline quartz and at other times amorphous silica. These two forms, as well as other polymorphs have substantially different solubilities as a function of temperature curves, although the solubility in general

Table 5-7. Analyses of a Typical Yellowstone Rhyolite (wt %) (Adapted from Allen and Day, 1935)

SiO_2	70.92
TiO_2	0.16
Al_2O_3	13.24
Fe_2O_3	3.59
FeO	0.66
MnO	0.14
MgO	0.23
CaO	1.42
Na_2O	4.28
K_2O	4.25
P_2O_5	0.18
H_2O	0.57

increases up to a temperature of 350°C where it levels out and then declines (see Fig. 5-2). A liter of water at a temperature of 200°C will dissolve between 0.2 g and 0.9 g of silicon dioxide depending upon the particular form in which it occurs.

Since the solubility of quartz in water is at its maximum at a temperature of around 335°C, 15° below 365°C, the critical temperature of water, water completely saturated with silica at this temperature will deposit it if either heated or cooled during its subterranean passage provided it remains long enough in one place. In this way, new channels can be opened up and old ones either enlarged or clogged, constantly modifying flow patterns. Large cavities may be dissolved out as at the extinct New Zealand geyser, Te Waro, to provide storage reservoirs for water. Or they may be closed off as is now slowly happening at White Dome, lying in Lower Geyser Basin, which is making itself extinct by sealing off its discharge vent with geyserite deposits. White Dome though quite active projects only a thin jet through a small orifice to a height ranging from 5 to 10 m for about 2 min from a huge sinter cone 4 m high and 20 by 30 m at its base. Intervals between eruptions are quite variable, ranging from 10 to 30 min.

Although cores of rocks greatly altered by the deposition of silica have been taken from drill holes as deep as 80 m at both Norris Basin and Upper Geyser Basin, the composition of the unaltered rhyolite is not very well known. The data available indicate that silica, originally dissolved at considerable depth along with bicarbonates, chlorides, and fluorides, seems to have been redeposited in the rhyolite lying at a few hundred meter depth.

A strong indication of rock interaction with the thermal waters at Yellowstone is the large amount of bicarbonates in the waters as contrasted with the very small amounts of carbon dioxide. The carbonate and bicarbonate ions are 0.7 times as abundant as the very prevalent chloride ions whereas almost no free carbon dioxide effuses from the springs. Comparisons of the relative abundances of the gases contained in the Yellowstone spring waters with those gases known to be given off by magma from which they are believed to come, indicate that the carbon dioxide is preferentially extracted by chemical reaction with rock, most likely rhyolites, to produce mica and clay minerals and bicarbonate and alkali ions. The carbon dioxide reaction is as follows:

$$CO_2 + H_2O + (NA,K) \text{ silicate} \rightarrow HCO_3^- + (NA^+, K^+) + H \text{ silicate}.$$

While the reaction of carbon dioxide with rhyolite is the most vigorous at considerable depth where temperatures are relatively high, sometimes it continues clear up to the surface, disintegrating surface rhyolite to form feldspar clays. Clay has also been found in core samples taken during drilling operations indicating that a similar reaction takes place underground as well. At Norris Basin, little carbon dioxide is converted to bicarbonate since the water rises through crystalline rock which is not as reactive as the glassy volcanic deposits found in other areas. The less chemically reactive nitrogen and methane escape from the geyser vent as pure gases.

The sulfate in the spring waters is derived from hydrogen sulfide gas which is

oxidized by atmospheric oxygen dissolved in meteoric water. A total of 15 ppm of the sulfate ion can be formed in this way from the oxygen dissolved in initially 5°C rain water. This is about the concentration observed at Upper, Midway, and Lower Geyser Basins and in the neutral waters at Norris. Acid waters relatively high in sulfuric acid result from surficial oxidation of hydrogen sulfide by the atmosphere. The combination of both acid water and chloride water in the Norris Basin is the result of mixing of shallow and deep waters in a near-surface reservoir.

Sulfate in the waters can result in unusual colorings. Kereru Geyser, located in the Whakarewarewa area, discharges pale green water and is surrounded by a black sinter mound. The colorings are believed to be due to compounds of iron, the green of the water caused by the presence of iron (ferrous) sulfate, and the black to iron sulfide (pyrite) derived from the sulfate by its reduction by hydrogen sulfide. The geyser erupts a fan-shaped jet at irregular intervals to a height of 10 m for about 30 s, followed by a series of 3 m splashes. Its plumbing system does not seem to be related to that of any of the other geysers in the area.

The chemical action of sulfuric acid on rocks is that which could be anticipated: it dissolves rocks like serpentine and disintegrates rocks like rhyolite leaving very fine particles, sometimes pure silica. The dull bleached-looking ground mass often associated with acid areas is residual material extensively pitted with angular cavities from which feldspar and quartz grains have been dissolved. The degree of reaction of sulfuric acid on the rock is extremely dependent on the amount of associated water, being the least reactive when diluted by abundant water.

At the high temperatures of underground geyser waters, potassium tends to replace sodium in the minerals from which rhyolite is formed and, as discussed earlier, provides a good geothermometer. As a result of the exchange, the subterranean rhyolite becomes potassium rich and spring water, sodium rich (Table 5-8). The sodium and potassium involved in the exchange are the chlorides, fluorides, and bicarbonates already dissolved in the water. While the exchanges are most often observed in alkaline springs as in the hot alkaline springs in New Zealand and Iceland, sulfate spring waters in which the rock is leached by

Table 5-8. Approximate Average Relation of Sodium to Potassium in Hot Springs and Geyser Waters (Atomic Ratio) (Adapted from Allen and Day, 1935)

Location	Na/K
Upper Geyser Basin	20
Beowawe, Nevada	12
New Zealand	30
Iceland	17−34
Steamboat Springs, Nevada	15

sulfuric acid also contain about the same quantities of sodium and potassium with potassium sometimes being the greater.

Analyses of thermally altered rock cores taken from drill holes at Norris Basin exhibit a ratio of sodium to potassium of 0.84, much lower than the 1.5 ratio normally found in rhyolite. Obviously, the sodium has gone into the spring waters and, indeed, it is found in excess in them. At the Upper Geyser Basin, the leaching exchange of sodium and potassium has progressed even further with the ratio of sodium to potassium in the altered rock cores being 0.47, suggesting either that the exchange has occurred faster at Upper Basin, or, more probable, that hydrothermal activity at Norris Basin is younger.

There also often occurs an exchange of the O^{16} and O^{18} isotopes between the thermal waters and the surrounding rocks, the exchange increasing the relative amount of O^{18} in the waters. This exchange occurs between course grained silicates and water and is significant only at temperatures above 300°C.

Mineral solution of quartz, and some oxygen isotopes exchange between water and rocks occur in the deeper aquifers. In the shallower aquifers, the following important chemical reactions and exchanges take place.

- Potassium-sodium ion exchange

$$(K^+ + Na\ rock = Na^4 + K\ rock)$$

- Rock alteration

$$(CO_2 + H_2O + Na\ rock = Na^T + HCO_3^- + H\ rock)$$

- Boiling and volatile loss

$$(2HCO_3^- = CO_{2\uparrow} + CO_3^- + H_2O)$$

- Mineral precipitation

$$(Ca^{++} + CO_3 = CaCO_3, etc.)$$

- Dilution

- Fluid-fluid reaction

$$(H_2S = 2O_2\ aq = SO_4^- + 2H^+)$$

- Surface oxidation

$$(Mn, Fe\ oxides\ ppt;\ H_2S \rightarrow sulfuric\ acid).$$

CHAPTER 6

Geyser Area Complexes

6.1 Fumaroles, Mud Pots, and Spouters

Fumarole, from the Latin word *fumus* meaning smoke, is a geologic feature that discharges only steam and other gases such as carbon dioxide and hydrogen sulfide at high temperature. They are commonly called steam vents and are found in volcanic regions, most often in lava, but also in geyser areas where they usually lie on higher ground than flowing springs. Water can be heard boiling violently deep within many of them.

Some fumaroles are spectacular. An 1867 description of the large, now flooded, New Zealand Karapiti Fumarole, located in the Orakei-Karako area, reads:

> With immense force, and amid loud hissing and booming the steam streams out of a circular hole in the loose masses of pumice stone at the foot of the hill. It is high pressure steam without a trace of any other gas, and bursts out through a small aperture in the depth of the circular hole in a somewhat slanting direction, with a sound like letting off steam from a huge boiler, and with such force, that branches of trees and fern bushes, which we flung into the jet of steam over the hole, were tossed into the air, twenty or thirty feet high.

Although no temperatures were recorded at Karapiti, the emerging steam was undoubtedly superheated which is often true of the more powerful fumaroles. Generally fumaroles are located in relatively dry areas and there are only a few, 10 to 15 in Yellowstone, with maximum temperatures of 138°C. There are numerous lower temperature ones, where steam at boiling temperature or lower escapes from fairly large vents. The more usual sight is where volcanic gases and steam bubble out of clusters of small vents covering many square meters. Hot spots can develop where the escaping gases impart a dancing motion to dry sandy soil; at other places, tiny jets of water will pulsate along hot rocks.

Black Growler, during the years 1925 to 1927, was the hottest and most powerful fumarole in Yellowstone. Steam at 138°C issued violently from a vent located at the bottom of a 13 m diameter, 1 m deep crater. There was a lower vent through which steam at a much lower temperature, 124°C to 125°C, began to flow. By 1931, the spring was inactive, although it has rejuvenated occasionally since then. Apparently action is related to changes in the amount of near-surface underground water. For example, an occasional thin sheet of water flowing along the crater floor would cause the temperature of the steam jet to fall. Following a period of wet weather when the crater partially filled with water, Black Growler would become a splashing, boiling hot spring until its excess water dried up and it reverted to a fumarole. While steam is by far the major component of the gas, small quantities of other gases are present to the extent indicated in Table 6-1.

Boulder Spring in Lower Geyser Basin lies at the foot of a sharp, 30 m high peak and is divided into two pools by a narrow barrier. Superheated steam shoots forth from one pool while the other performs as a perpetual fountain-type spouter, occasionally acting like a geyser and throwing up fans of water a few meters in diameter to heights as great as 2 m.

Mud pots, which are essentially nearly waterless hot springs, are among the most fascinating and interesting features of geothermal areas. The available water is mixed through with fine undissolved mineral matter, mostly clays, to form a viscous wet mud of various consistencies from very soupy to nearly hard-baked material. They occur generally in areas where the considerable hydrogen sulfide present in acid spring waters decomposes the surface rocks. When water flow is high, the fine residual mineral matter is washed away; otherwise the water in the pool is either turbid or almost totally saturated with mud. The mud commonly is gray, black, white, or cream-colored; colored mudpots, "paint pots", are pink or bright red from the presence of iron oxides. The colors are basically due to a shortage of sulfur since an abundant amount would transform the iron oxides into pyrite, a gray-colored mineral. When the mud is quite viscous, large intact blobs of it may be thrown out and piled up as a cone or mound, generating a "mud volcano". Hard-baked, nearly extinct mud volcanoes are fairly common. The little steam which is discharged from a small opening at the top suggests that the vent has been mostly sealed off underground.

Mud pots occur singly and in groups and are widely scattered throughout most

Table 6-1. Composition of Gases Discharged from Black Growler (Adapted from Allen and Day, 1935)

Gas	Percent
Steam	99.600
Carbon Dioxide	0.386
Hydrogen Sulfide	0.010
Hydrogen	0.002
Methane	neg
Nitrogen-Argon	0.002

geothermal areas, especially in Yellowstone where they fill many areas a few square meters to as much as a hectare in extent. Individual pots range widely in size from small holes in the ground to kettle-like and funnel-shaped basins 10 m in diameter and 5 m deep. In some areas, the pots form a nearly continuous sheet, while in others they are separated by baked, dry ground.

As puffs of steam burst through the mud in active areas, concentric rings of expanding waves frequently form on the surface. The action in stiffer muds may be quite spectacular when the sudden explosive outbursts of steam generate showers of hot mud or produce knobs, lilies, and other fantastic shapes. The surface patterns of Ngapuna Tokatoru mud pot, located close to Puarenga Geyser in the Whakarewarewa area changes seasonally. During the dry summer when the mud is thick, small volcanoes build up; during the wet winters, the mud is thinner and concentric patterns and "lily and rose" shapes form. The visibility of its surface features is considerably enhanced by a natural thin cover of black, oily graphite.

Microscopic examination of the Norris Basin muds which are probably fairly typical show them to be composed of a mixture of partially altered rhyolite, large amounts of quartz, a little opal, with a liberal amount of clay mineral containing firm separate grains of kaolinite. The finely divided clay and opal is similar to the sediment found in hot springs. Both are almost certainly derived by the decomposition of rhyolite with sulfuric acid and hence mud pots are classed chemically as acid springs. A chemical analysis of a typical Norris Basin mud pot is listed in Table 6-2. As might be expected in an acid water, the dissolved salts are mostly sulfates, 474 ppm, rather than carbonates and chlorides.

A perpetual spouter develops when successive eruptions of a geyser coalesce, almost completely erasing the quiet periods between them. The rate of discharge of water and steam from the spring continues to pulsate, generating substantial variations in heights and rates of jetting. Some geysering hot springs, such as

Table 6-2. Ionic Composition of Norris Basin Mud Pot Water (ppm) (Adapted from Allen and Day, 1935)

Chemical Ion	(ppm)
H	0.25
NH_4	107
Na	25
K	50
Mg	3
Ca	13
Al	tr
Fe^{++}	13
Cl	45
SO_4	474
SiO_2	236

Imperial in Yellowstone and Yosimoto in Japan, alternate between being a geyser or a spouter.

Imperial Geyser, for example, although now a continuous spouter, was famous as a geyser in the late 1920s when it suddenly appeared and functioned spectacularly for a couple of years. It is located about 8 km up Fairy Creek, south of Lower Geyser Basin. Starting out as a slender 8 m high nearly continuous jet of water in 1927, by 1928 it had become a powerful geyser projecting almost continuously 10 m wide splashes to heights ranging from 15 to 25 m for about 3 hr. Quiet intervals during this time never exceeded 5 s. After the activity abruptly ceased, the now smooth water would begin to disappear, and in half an hour, having dropped 2 m in level, be completely gone, exposing a 50 by 100 cm wide by 6 m long fissure. On the uncovered crater floor, water remaining in numerous small depressions would boil vigorously and steam would escape through the wet sand that covered the more elevated areas for some time after the eruption stopped. The sustained eruptions occurred about every 12 hr.

The geyser's basin changed considerably during this time. Originally it was a 25 m diameter shallow circular pool with a 1 to 2 m high vertical wall. After the large eruptions began it gradually increased in size to a 26 by 35 m crater in 1928 and to a 35 by 45 m one by 1929. The fissure in the floor, from which the geyser played, bifurcated, generating a 5 m wide by 10 m long branch, and extended itself to 3 m wide and 8 m long.

During 1929, the periodic eruptions became continuous, much less violent, and have remained more or less so since then.

Papakura Geyser, perhaps the most active spring at Whakarewarewa, is a typical spouter. It is usually in continual turmoil, shooting jets of steam and water in all directions up to heights of 5 m.

The actions of perpetual spouters are easy to account for. They are hot springs through which steam is pushing its way upward, pulsations in the jets being caused by the resistance offered by the fluid to the passage of this steam. The steam in most cases is superheated. Steady Geyser, an unusually powerful spouter in Yellowstone, propels a very large jet of water to a height of 10 m while at the same time maintaining an almost boiling temperature. The power of the jet pulsates somewhat, even occasionally dropping to the ground. Other less vigorous spouters do not function nearly as regularly.

Spouters as well as fumaroles frequently occur in ground where no siliceous sinter has been deposited. Thus there probably exist no sinter-lined underground chambers tightly enough sealed to hold steam. The occasional geyser that develops in such an area is usually not long-lived.

6.2 Nonerupting Hot Springs; Boiling Springs

A very great number of nonerupting hot springs, generally referred to as intermittent or pulsating springs, exhibit a periodicity in their action. Two rather distinct types of actions take place in such springs: intermittent rise and fall of the water level leading to intermittent overflow; and intermittent mild eruptions. The

changes in water level result from the gradual expansion and contraction of steam bubbles probably mostly located within underground chambers; the eruptions are precipitated by sudden subsurface boiling. Either one or the other or both of these actions may take place in a single hot spring. There is not a clear and distinct difference between an erupting intermittent spring and a geyser with regard to the nature of the eruptions. But those springs with well defined and energetic eruptions are usually termed geysers.

Many intermittent springs are found in Iceland especially near Reykir in Olfus. They exhibit several actions. One small spring having a 15 cm diameter, 20 cm deep basin boils but never erupts. Following violent boiling, the basin empties, followed by slow filling with gently boiling water, with the boiling gradually becoming more violent until the basin empties again, the whole process taking about an hour. Another small spring boils violently most of the time but every 12 min stops for about a minute. Two others swell regularly for about 40 min, after which the water retreats for 25 min before starting to swell again.

Some intermittent springs act almost like geysers, differing only in the heights of their eruptions. At Beowawe, there are some springs that intermittently erupt to a height of only a few centimeters. The basin of the largest spring of this type is a 2 m diameter, sinter-lined funnel extending to a depth of more than 3 m. Its few centimeter high rim is breached in one place. It takes about 20 min for a cycle to run its course. First, the level of the 90°C water which stands about 30 cm below the rim's outlet, begins to rise as approximately 10 cm diameter gas bubbles start appearing at the opening at the bottom of the funnel and floating to the surface. By the time the water begins overflowing, about 10 min later, its temperature has gone up gradually to 94°C. Bubbles now appear much more rapidly, finally causing the water to erupt to a height of about 12 cm. In the meantime, the steady overflow of water is about 20 liters/min. After the eruption, everything returns gradually to its starting condition, the water level falls, the amount of gas discharged decreases, and the water temperature drops to 89°C.

Several of the Japanese hot springs are classed as intermittent springs although also exhibiting geyser-like eruptions. The discharge and water levels in these springs fluctuate more or less regularly.

Intermittent flow of this kind seems to be caused by certain instabilities that develop when water, steam, and gases mix. Basically two situations are possible: foam, the dispersal of gas bubbles in water; and mist, the dispersal of water droplets in a gas-filled space. Which develops in a spring depends upon the relative volumes of gas and water present, and upon the velocity of flow in the discharge vent. Figure 6-1 defines the conditions under which each will exist. Mist will exist when the volume of gas is relatively high; and foam, when the volume of water is high. Under certain conditions, neither can exist and intermittent flow takes over in the following manner. Assume that the fluid exists as a foam, corresponding to a point A in Fig. 6-1, at some point down within the central tube of the spring. As it begins to rise, the hydrostatic pressure decreases and accordingly the volume of gas increases soon reaching a value corresponding to point B. Foam cannot exist if the amount of gas were to increase. On the other hand, the amount of gas is not sufficient to form a mist. The only recourse is to

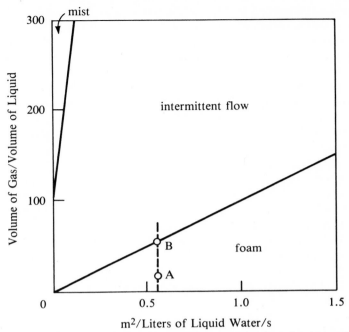

Figure 6-1. Conditions for the formation of foam and mist in a flowing well. (After Barth, 1950.)

expel gas but for this to happen the flow of water must temporarily stop which it apparently does, decreasing the discharge from the spring. The hydrostatic pressure will change due to the loss of gas creating an unstable situation that under suitable conditions leads to regular intermittent flow. It should be noted that these are essentially the same events initiating geyser action and are probably responsible for so-called "unsuccessful" eruptions or "floths" as referred to by the Icelanders.

The floths, which are essentially subterranean eruptions, are much more powerful than similar eruptions in large hot water basins as a consequence of the choking effect of the narrower geyser channels. Stori Geyser, Iceland, is such a one. In addition to its large surface eruptions, it has regular subterranean eruptions. Both kinds are preceded by underground detonations but when there is no geysering, there is only a violent surge of steam and water, causing the surface of the water in the basin to rise up into a conical shaped mound. Somewhat similar floths occur in geysers at Beowawe and Yellowstone.

6.3 Temperature Regimes within Geysers

Thousands of temperature measurements have been made down within hot springs and geysers, usually by means of thermistors, electrical resistance thermometers, or maximum temperature mercury thermometers lowered on a wire.

The characteristics of the temperature regimes vary widely from geyser to geyser and from time to time within any one geyser.

The temperature is ordinarily almost constant with both time and depth in relatively open pools where active circulation thoroughly mixes the water and tends to equalize the temperature. Temperature measurements made a few years ago indicate that this is true in the large pools at Artemisia (Fig. 6-2a) and at Great Fountain (Fig. 6-2b). In both cases, geyser action seems to be initiated by injection of very hot water from small holes in the bottom of the pool.

The pool at Great Fountain appears to be fed with hot water flowing through a vent in the bottom of the 2 m deep pool. At the surface, the pool is essentially at the local boiling point, 93°C; 6 m down it fluctuates between 95°C and 97°C. A large convection cell develops within the pool with cooler water descending along the wall of the tube, moving inward as it approaches the bottom, becoming heated at the center by hot water flowing from the vent, and finally rising to the surface being cooled as it does so.

An eruption does not completely empty the pool, only lowering its water level about 3 m. The voluminous quantity of water ejected during the geyser's approximately 1 hr eruption indicates the existence of other hidden reservoirs. The temperatures within the pool, which stay essentially steady for several hours,

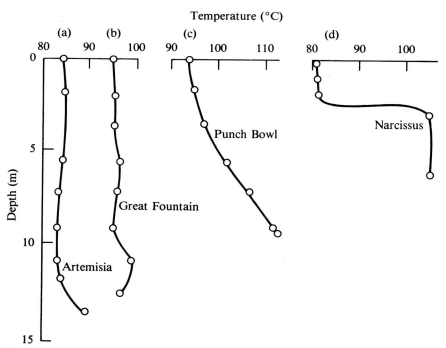

Figure 6-2. Temperature-depth curves for several types geyser reservoirs: (a) Artemisia, (b) Great Fountain, (c) Punch Bowl, (d) Narcissus, (e) Giantess, (f) The Great Geysir, and (g) Beehive. (Adapted from Allen and Day, 1935, except for (d) from Rinehart, 1970a, and (f) from Birch and Kennedy, 1972.)

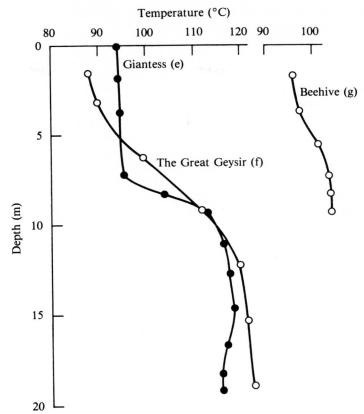

Figure 6-2. *(continued)*

suddenly increase just before an eruption about 10°C, up to 107°C. Surface boiling becomes much more violent, a large amount of overflow occurs, and the geyser begins to erupt. Such large scale heating could only occur if large amounts of hot water or steam are suddenly forced into the pool from below.

When the shape of the basin impedes free convection, temperatures generally build up near the bottom, resulting in a monotonic increase of temperature with depth. Punch Bowl, Yellowstone (Fig. 6-2c) is such a spring. It boils and overflows almost continuously with periodic surges. Water in its narrow basin is heated from the bottom by the influx of steam and hot water through a narrow opening.

In more complex cases, such as Narcissus having two or more connected reservoirs, free circulation can take place in each but is difficult between them. The temperature-depth curve will have two displaced vertical segments connected by a horizontal segment at the depth corresponding to the depth of the narrow connective constrictions (Fig. 6-2d).

A similar type of temperature-depth curve is found when a hotter, side reser-

voir or inlet is injecting water or steam into the main reservoir at a point part-way up its side as seems to happen at about the 8 m level in Giantess (Fig. 6-2e).

A somewhat modified form of the curve will develop when the constriction is not especially severe so that a mixing cell develops as formerly happened at the Great Geysir when it was active. Three distinct regions can be seen in Fig. 6-2f. The temperature down to 6 m, a distance about twice the diameter of the tube, remained nearly constant: between 80°C and 90°C. At depths of 12 to 18 m, the deepest the geyser could be probed, the temperature stayed constant at about 120°C. In between, however, from the 6 to 12 m depth, violent mixing took place early into the eruption cycle with the temperature fluctuating widely and bubbles of steam forming under the reduced ambient pressure. At first these steam bubbles collapsed, although their increased volume temporarily raised the water level in the geyser basin. Later on in the cycle as the temperature of the water rose, they did not collapse but initiated the eruption.

One of the simplest temperature regimes (Fig. 6-2g) is that found at Beehive, a gradual rise in temperature with increasing depth down to considerable depth after which the temperature remains nearly constant. Presumably geysers where such a distribution is found have only a single reservoir tapped from above by a narrow tube venting at the surface.

The temperature regimes in Crump Geyser, a drilled geysering well located in Lake County, Oregon, have been studied in considerable detail. The geyser throws up a single column of water to a height of about 30 to 40 m on an 8 to 10 hr schedule. The tube fills quickly after each eruption. The geologic setting (Fig. 6-3) is fully documented from cores taken during the drilling operation. The well is 35 m deep, with the upper 17 m in alluvial that is underlain by a layer of sand saturated with hot water and extending from a depth of 17.7 to 19.8 m. The water-bearing sand is the source of heat and water. The lower 15 m of the hole is in solid lava.

Temperature measurements indicate that hot water at a temperature of about 120°C enters the well at about the 18 m level. It first fills the lower part of the well and then begins to flow upward, cooling as it does so. When the well becomes full, circulation develops about the 7.5 m level, further heating the water in the uppermost part of the well. As the column of water becomes hotter, temperatures begin more and more to exceed the ambient boiling temperature, especially at about 7.5 m level. Finally the column goes critical and an eruption occurs.

Two distributions of temperature as a function of depth are plotted in Fig. 6-4, one taken an hour or so after an eruption when the tube is filled with water and the second taken 8 or 9 hr later, just before an eruption. Curve 1 looks about as to be expected in a narrow single tube fed part way down by lateral flow of hot water. It lies below the boiling point curve all the way to the bottom of the well, becoming almost tangent to it at 7.5 m where the curve abruptly bends downward becoming essentially vertical. The continuous temperature-time recordings at the 7 m level show temperatures fluctuating 2°C to 3°C. Thus Curve 1 must be piercing the boiling point curve from time to time producing steam bubbles. The

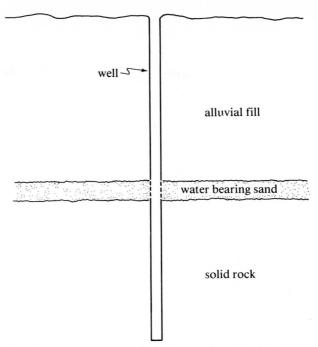

Figure 6-3. Geologic cross section of Crump geysering well.

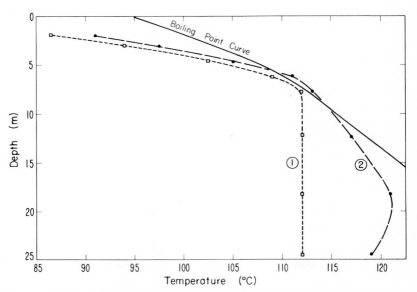

Figure 6-4. Down-the-well temperature profiles in Crump: (1) an hour or so after an eruption, (2) just before an eruption.

very sharp curvature in Curve 1 near the point of tangency is attributable to the fact that water flashing into steam leaves the residual water much cooler. During the early stages of the interval between eruptions, the rising steam bubbles quickly cool and collapse in the upper water, further warming it. The fairly high fluctuations in temperature in the waters lying above about 3 m, 3°C or 4°C, indicate considerable circulation there.

As time passes, the segment of the temperature-depth curve below 7.5 m begins to swing to the right, toward increasing temperatures, with the water above the 7.5 m level heating up as well. Slowly the whole curve moves to the right (Curve 2), penetrating further and further into the boiling regime as a temperature maximum develops at the 17.7 and 19.8 m level near the inflow of water.

The column goes critical when the upper water no longer has the heat absorbing capacity to condense the steam coming up from below and an eruption occurs.

Many temperature measurements have been taken within Old Faithful. Although in general temperature increases with depth, it varies substantially in the same manner at all depths down as far as the 16 m level. A typical upper level curve, taken at a depth of 23 m, is shown in Fig. 6-5. Following an eruption, the temperature remains constant at about the ambient boiling point, 93°C, indicating that the incoming water filling the tube has not yet filled the tube up to the 23 m level, the temperature sensor simply being bathed in a cloud of steam rising from below. The length of time it takes to fill the tube and engulf the sensor correlates closely with the time to the next eruption. Soon after the water reaches this level, the temperature rises quickly to 98°C and then more gradually to about 105°C taking 10 to 20 min to do so. After this it begins to cool, undoubtedly as a result of convection, lowering in a few minutes to a temperature of 95°C or 96°C. Two or three minutes before an eruption starts, it again rises and after the eruption starts, goes up still higher to temperatures ranging between 107°C and 112°C. As the eruption proceeds, the temperature decreases almost linearly with time, dropping to 93°C when the eruption is complete.

The character of the variation of temperature with time (Fig. 6-6) changes significantly between the 60 m and 90 m level. The water at 90 m remains at a temperature of about 105°C or 106°C during most of the interval. The rise in temperature preceding the eruption begins 5 to 10 min ahead of the eruption,

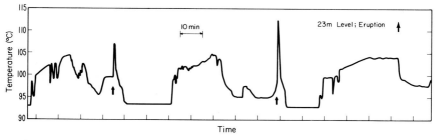

Figure 6-5. Temperature as a function of time at the 23 m level in Old Faithful.

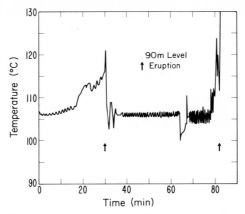

Figure 6-6. Temperature as a function of time at the 90 m level in Old Faithful.

much before the 2 to 3 min observed at the 10 to 60 m depths. At eruption, the temperature has reached 120°C, 8°C to 10°C hotter than at the levels closer to the surface.

Fairly regular upward and downward movement of water masses continually occur, especially from 90 m downward, but occasionally in the 15 to 60 m range of depths. These movements appear on the temperature-time curves as rapid 1°C to 2°C fluctuations (Fig. 6-6) occurring at about 1 min time intervals. The temperature at the 30 m depth level drops 2°C to 3°C every time a steam bubble breaks at the surface.

Very deep within Old Faithful, 175 m down, the deepest it has been probed, the variation of temperature with time is substantially different (Fig. 6-7). For about 20 min following an eruption, the temperature rises gradually, at a rate of

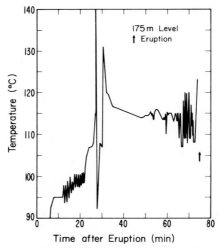

Figure 6-7. Temperature as a function of time at the 175 m level in Old Faithful.

about 0.5°C/min, except for oscillations produced by upward and downward convection and a short period immediately following the eruption when the temperature remains constant at the boiling point, 93°C. The steady rise is interrupted by a large surge of extremely hot water, hotter than 140°C, almost as if a subterranean eruption occurred. This is followed by a surge of cooler water, which, in turn, is followed by another surge of extremely hot, but not quite as hot water; the rapid circulation takes place in about 5 min. The temperature finally levels off at 115°C, except for periodic excursions of 5°C. The temperature does not rise or fall either as eruption time approaches or while the eruption is occurring.

Temperature as a function of depth at various times before an eruption are shown in Fig. 6-8. The curves are fairly complex due mostly to intermittent

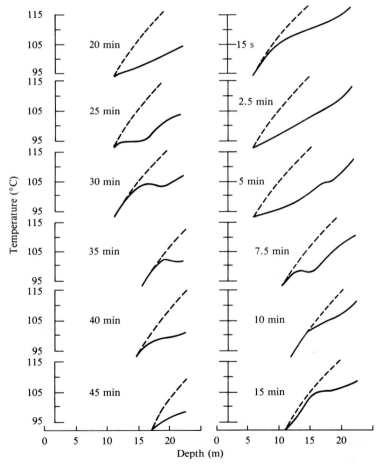

Figure 6-8. Profiles of temperature versus depth at Old Faithful at various times before an eruption. Time indicated by number beside curve. (After Birch and Kennedy, 1972.)

mixing as instabilities develop. These are evident on the temperature records taken at fixed depths (Figs. 6-5, 6-6, and 6-7). Clearly these curves support the notion that the eruption is initiated in the uppermost 5 m of the water column, the region where the water temperature eventually exceeds the ambient boiling point. Except at these very shallow depths, the temperatures between eruptions are considerably less than the ambient boilding point. Immediately before an eruption the temperatures near the top increase very rapidly, bringing the temperature-depth curve and the boiling point curve into coincidence in this region, leading to an eruption as the water begins to boil.

6.4 Interactions among Geysers

Most major geysers seem to act independently of one another even when closely located. Even if the siliceous sinter, as at Upper Geyser Basin, forms a continuous thick surface pad, the plumbing system of each geyser is isolated from that of any other. A striking example of such isolation was evident in the last century during the explosive and extremely violent eruptions of Excelsior Geyser. Nearby Prismatic Lake, only a few meters distant, remained perfectly calm. Further evidence of their isolation are analyses of the chemical compositions of waters taken from geysers in the same basin. There are substantial differences. For example, Giantess, Beehive, and Artemisia Geysers have the following sodium-potassium ratios, respectively: 362/20, 322/18, and 422/16.

While the remarkable feature of many geysers is their complete independence of action, there are some cases where several geysers and hot springs interact to regulate each other's activities. Especially those exhibiting great irregularity have been shown to be subterraneously connected with neighboring hot springs and geysers enabling the action of one to influence that of another and permitting redistribution of energy from time to time. The rise of the water level in one spring may be accompanied by a recession in another; one regularly erupting geyser may stop erupting and a nearby dormant geyser may become active again; or the eruption of one geyser will trigger or quench that of another.

Simultaneous visual observations of the Grand group of geysers, Grand, Turban, Vent, Triplet, and Rift, all within 35 m of one another, hint at interconnected plumbing systems. Turban and Vent, both much smaller geysers than Grand, play in synchronism with the spouting spring. Triplet and Rift Geysers seem to exercise considerable control of Grand's activity, mainly by bleeding away energy. During the years when Grand is relatively inactive, erupting only every few days or weeks, Triplet and Rift are quite active; and vice versa, when Grand is very active, erupting two or three times a day, Triplet and Rift are relatively inactive.

Triggering, producing a sort of chain reaction, has been common among the Fountain group of geysers in Lower Basin. Clepsydra and Fountain at one time always erupted in series on the heels of an eruption of Morning Geyser, apparently as a consequence of water from Morning's 45 min eruption flowing in and

filling their basins. Clepsydra began playing through its four vents about 8 hr after Morning stopped and continued to play for 3 to 6 hr. Fountain erupted for about 45 min, 6 to 8 hr later.

Although surface manifestations may point to the existence of subterranean interconnections between geysers of certain groups, more definitive techniques have been used to explore the matter further. One of the most successful has been the use of sodium iodide as a tracer. It is especially suitable because it is initially absent from geyser waters, it can be detected in as small amounts as 0.1 ppm, it cannot be removed by absorption or ion exchange, it is relatively inexpensive, it presents no health hazards, and it is very soluble in water.

The Daisy group in Upper Geyser Basin, consisting of Comet, Splendid, and Daisy geysers, Bonita and Brilliant pools, a small unnamed geyser 13 m south of Brilliant, and two pools near Splendid all lie within an area of 13 by 45 m. Surface features point to interconnections and there occur sympathetic geyser eruptions and simultaneous changes in water levels in the pools. The stratagem used to study the possible existence of interconnections was to add 3 kg of sodium iodide to Splendid and then at intervals to sample the water in the other springs to observe the concentration of sodium iodide in each. An interconnection certainly does exist and in addition it was found that an underground reservoir of at least 260,000 liters of water exists, with most of the discharge from it occurring below ground.

Apparently there are close underground interconnections between several of the geysers at Whakarewarewa. Pohutu, the most famous, generally erupts to a sustained height of about 18 m, although starting its eruption much higher. An eruption of Pohutu is always accompanied by one at Prince of Wales Feathers, usually preceding it with a feeble eruption that increases in strength as Pohutu begins its eruption, and reaching a steady height of 12 m. Nearby Waikorohihi frequently erupts a 5 m pulsating jet, and occasionally a higher continuous jet but an eruption of Pohutu stops its action. The waters in Te Horu always rise to herald an eruption of Pohutu, most frequently boiling vigorously and overflowing, or more rarely, erupting to a 15 m height. Fluorescein dye added to Wairoa and Te Horu established that a very direct, shallow connection exists between Te Horu, Pohutu, and Prince of Wales Feathers and that whereas a connection exists between Te Horu and Waikorohihi, it is indirect and separate from the connection between Te Horu and Wairoa.

An arrangement of surface openings, underground channels, and reservoirs which can account for the above geyser and pool behavior and distribution of the flourescein dye is illustrated in cross section in Fig. 6-9. The distribution of water in the geyser system immediately after an eruption of Pohutu (P) is shown in Fig. 6-9a. The position of the water L_2 depends upon the length of the eruption just finished. After the occasional long eruption it may fall as low as L_1. The water in reservoirs C_1 and C_2 is superheated but is below boiling in C_3.

Shortly after the eruption, the outlets begin to fill with water whose temperature and that in C_2 and C_3 are increasing as a result of mixing with steam and gas bubbles released from below which enter the system through side channels, S.

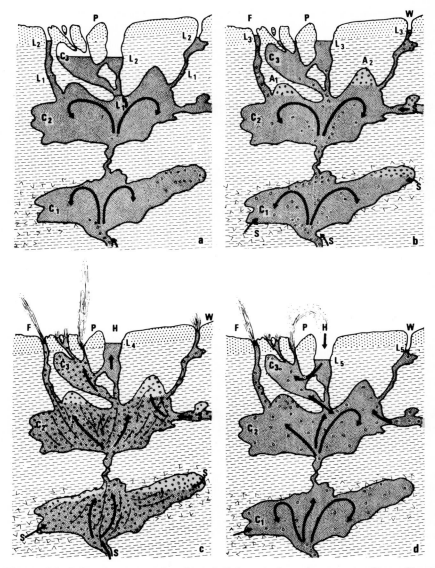

Figure 6-9. Inferred cross section through Pohutu and nearby geysers. (From Lloyd, 1975.)

Some bubbles collapse suddenly, producing audible shocks in the system. As the water warms up, the bubbles no longer condense and the gas is trapped in domes A_1 and A_2 of reservoir C_2 (Fig. 6-9b). Some gas escapes up the vent tubes of Waikorohihi (W) and Prince of Wales Feathers (F), reducing the hydrostatic pressure exerted at A_1 and A_2 by lowering the density of the overlying water allowing more gas to escape from A_1 and A_2 and precipitating eruptions of both Waikorohihi and Prince of Wales Feathers.

The release of pressure resulting from the eruptions greatly accelerates bubble formation, raising the level of the Te Horu (H) column from L_3 to L_4 and setting Pohutu into eruption as a consequence of the water in C_2 flashing into steam under the just reduced ambient pressure. Now the whole system is in violent action (Fig. 6-9c). Since water discharges faster than it is supplied, water levels will lower and the transformation to steam of much of it will change the composition of the residual water. These changes in the composition of water ejected by Pohutu during an eruption are listed in Table 5.4. The nonvolatile chloride increases its concentration in C_1, C_2, and C_3 whereas the volatile ammonia, NH_4, is carried along with the steam and discharged. The concentrations of chemicals in the waters are seen to increase as Pohutu's eruption progresses and the ratio of chlorides to ammonia decreases radically.

Eventually so much water is lost that the water level L_4 falls substantially to L_5, far enough below the surface vents of Pohutu and Waikorohihi to quench the eruptions (Fig. 6-9d).

Many other tracers for underground study are available. Stable isotopes are excellent for use in natural waters but they are expensive and require a mass spectrometer for analysis. The same holds true for radioactive isotopes and in addition these latter might possibly present a health hazard.

CHAPTER 7

Environmental Aspects of Geysers

7.1 General

Hot springs and geysers add abnormal amounts of heat energy, mineral matter, and water to highly localized regions of a normally balanced ecosystem. As a consequence, these areas develop local anomalies in their biologic and geologic features and sometimes even modify the atmospheric environment. The extreme temperatures, violent processes, and unusual nature of such areas make them both beneficial and hazardous to plants and animals, including man.

Generally animals, sensing hot and dangerously fragile ground, avoid it. Occasionally accidents do occur as evidenced by bleached bones at the bottoms of some hot pools. One of the greatest hazards, especially to humans, is due to the fact that many springs and geysers have rims and surrounding sinter platforms that appear firm and safe but are in fact friable, dangerous, and underlain with scalding water. Although those geyser areas frequented by tourists are usually equipped with warning notices, boardwalks for viewing and are well fenced, occasionally a spectator is painfully and sometimes seriously injured when ground gives way under him or he inadvertently steps into a pool of hot water.

While the behavior of many geysers is predictable, that of many others is not. Occasionally a geyser will erupt at a completely unexpected time or with unanticipated violence. When this happens spectators may be taken unaware and subjected to minor or serious injury.

As hot spring and geyser activity changes, new areas will become hotter; others will cool. Such changes can drastically affect plant life in the area. In some forested areas in Yellowstone, trees died row by row following the Hebgen Lake earthquake as the extent of the hot areas gradually spread. Such devastation is a common sight in Yellowstone (Fig. 7-1). The usual area exhibits a large amount of dead, white, petrified-like trees and wood, some still standing as ghost

Figure 7-1. Devastated forest area near Biscuit Geyser Basin, Yellowstone.

forests, others as logs lying on the ground, and some as branches and sections of trunks completely embedded in masses of siliceous sinter or partly embedded in the silica border of a thermal spring. Siliceous material is deposited in the wood as it takes up and then evaporates water heavily laden with silica. True petrification, the replacement of wood by opal, does not take place. Rather the silica precipitates and forms thin layers along cell walls and ducts, sometimes even filling them. In this way a skeleton is formed which preserves the minutest details of the wood cells and fibers.

Although the average mineral content in geyser waters is relatively low, great quantities of material can be accumulated or transported over long periods of time. The bulk of the geyser and hot spring waters flows away and the remainder forms basins, cones, terraces, platforms, and stream channels. Some is deposited on trees and rocks as impinging sprays evaporate or in cold weather freeze into magnificent mantles of ice that leave a deposit of silica when they melt.

7.2 Mineral Deposition

Silica is the surface-deposited material most closely associated with geysers that sit as clear pools or raised cones in the center of a broad, barren, pale sheet of intact or broken up siliceous sinter (Fig. 7-2). The barrenness may be relieved by brightly colored patches of algae, especially in small depressions and drainage channels. The sheets are often very extensive, beginning as the lining of the geyser's pool and vent, running under the overflow and drainage channels and extending to the bottom of nearby rivers or lakes.

The dissolved mineral matter in geyser and hot spring waters ranges from one to over ten grams per liter. Analyses of the river waters at Yellowstone have indicated that they carry out about a total of 352,000 kg of mineral matter per 24 hour day.

In addition to the slow deposition of dissolved minerals, an occasional violent steam explosion occurs and strews irregular blocks of sinter about along with rocks and fine sands.

Close to the pool or cone where the silica is more or less constantly wet, dense glassy opal forms into ring-like borders up to as wide as 50 m. A spring may deposit a thin sheet of sinter on its surface, resembling a sheet of ice extending out from the bank (Fig. 7-3). Depending on the depositional environment, the opal is found in many different shapes: like cauliflower, warty-like protuberances (Fig. 7-4), glazed laminated surfaces decorated with pearly white beads, or

Figure 7-2. Sinter cone of Lone Star Geyser, Yellowstone.

Figure 7-3. Sinter incrustation surrounding small hot pool at Yellowstone.

sinuous snake-skin-like patterns. "Geyser eggs" found around geysers are particularly fascinating (Fig. 7-5). These are small rocks, varying from shot size to the size of a hen's egg, which have been completely coated with one or more layers of precipitated opal.

Glassy opal, which is heavily hydrolized silica, dries out and coats itself with a white, opaque powdery material when it is found some distance from the pool or vent where it is relatively dry most of the time. Old, dried-out sinter is quite fragile and brittle. It breaks up easily by the weathering process into chips, flakes, and coarse sand. Sometimes as a result of changes in a geyser's activity, the pieces can be recemented together into a competent mass by newly deposited sinter.

The chemical compositions of two sinters, one from near Daisy Geyser, and the other from near Pohutu, are listed in Table 7-1. The sinters are primarily opal but at Pohutu also contain some chalcedony, jasper, agate, and crystalline quartz

Figure 7-4. Abrasive, sponge-like sinter in opening of Narcissus Geyser.

Figure 7-5. Geyserite eggs surrounding vent of Narcissus Geyser.

Table 7-1. Chemical Composition of Sinters (wt %) (Adapted from Lloyd, 1975; Allen and Day, 1935)

Mineral	From Near Pohutu	From Near Daisy
SiO_2	85.58	89.75
Al_2O_3	0.74	0.38
Fe_2O_3	0.09	0.06
MgO	0.01	0.14
CaO	0.64	0.56
Na_2O	0.58	0.26
K_2O	0.28	0.12
TiO_2	0.03	
ZrO_2	tr	
P_2O_5	0.01	
S	0.09	
MnO	0.02	
BaO	0.03	
Cl		0.03
SO_3		0.05
Combined water and organic matter	6.17	
Water lost at 104°C to 105°C	5.50	9.36

as well as trace amounts of pyrite, gold, and silver. The higher water content of the Yellowstone sinter as compared with that of New Zealand, probably reflects the colder and wetter climate at Yellowstone.

Silica is also deposited below ground level to form linings on already existing vent tubes, subterranean channels, and reservoirs, to build new tubes and channels, and to extend others. Some scientists believe that without the depositional building and modification of channels and reservoirs, geysers would not exist at all, although continued deposition does not guarantee geyser action. During the period from 1915 to 1935, when the Great Geysir was not erupting, a 10 cm thick layer of geyserite was deposited on the bottom of its submerged bowl. While the plumbing of the smaller geysers may not extend below the 6 m thick sinter deposit of the Yellowstone area, that of the larger ones such as Old Faithful extend a few hundred meters, suggesting that deposition of sinter may take place at considerable depth.

The characters of the deposits of Norris Basin range widely among geysers and springs, reflecting an equally wide variation in the natures of their respective waters. In certain springs opaline sinter readily forms under water and in many small pools a heavy white fresh coating of sinter covers twigs and pebbles. At Porcelain Terrace, the most active of the opaline depositing areas, a 32 mm thick layer of it accumulated in a 10 month period. Occasionally an acid spring at

Norris will precipitate sinter, which bears out the observation that the acid waters extend only to a depth of about 3 m before they revert to silica-bearing alkaline waters. In general, however, the alkaline springs precipitate sinter abundantly whereas the acid springs precipitate little if any at all. The rate of deposition is variable and does not seem to be related to the amount of silica in the water of a spring or to the presence or absence of organisms such as algae, bacteria, or diatoms.

The deposition of silica is often attributed to the fact that silica is more soluble in hot water than cold and that precipitation occurs as the water cools but that is certainly not the complete explanation. In some experiments, containers full of hot geyser water containing as much as 700 ppm of silica, an amount greatly in excess of its solubility at room temperature, were found not to precipitate out any silica even after sitting for weeks. Evaporation can account for a large part but not all of the deposition. Freezing separates out the opal, but freezing does not appear to be a likely factor in warm New Zealand where many sinter cones and pads have built up. Experiments in which silica is deposited from dilute solutions onto submerged objects such as small boards and rocks, leave no doubt silica can and does deposit out but the factors controlling its deposition not only varies from spring to spring but also does not always remain constant within a single spring. Only one general statement is possible: the very highest rates of deposition are found in the areas having the highest concentrations of silica in the spring waters.

Free sulfur is being deposited at both Norris Basin and at Whakarewarewa in cracks, cavities, and porous material. Occurring sometimes in both chloride and sulfate areas, generally around fumaroles, it is probably a secondary oxidation product of hydrogen sulfide.

Salt deposits are also commonly found around neutral springs at Norris and Whakarewarewa. The salts which are deliquescent and homogenous, are usually sulfates of the metals commonly found in rocks, with potassium present in greater quantities than sodium.

7.3 Plant and Animal Life

Warmed by the hot waters and fed by their gaseous and mineral constituents, luxurious vegetation is often rampant near hot springs, especially those whose waters contain nitrogen. Quite the opposite is true near sulfated springs and fumaroles. In Yellowstone, in spring and early summer, the small bright yellow monkey flower and the lily-like yellow fritillary are prolific. The vegetation can almost cover the vent when the water temperature in the spring is low enough.

Hot water can kill existing plants when a new hot spring comes into existence or the overflow from an existing spring changes the course of its channel.

Brilliant colors, brown, white, yellow, salmon-red, and blue-green, often surround and line the hot pools and geyser run-off channels, sometimes by reflection coloring the rising steam. These colors are produced by living organisms that have adapted themselves to the extreme temperatures of the hot

spring waters as well as their pH. Several distinctive groups of both plants and animals are found around alkaline springs depending upon temperature (see Table 7-2).

Whereas the pH of the water of a particular spring remains nearly constant, its temperature does not, cooling as it flows down its drainage channel or spreads out toward shallower edges. Those organisms best adapted for the highest temperature flourish nearest the source whereas those adapted to the lower temperature grow further down the drainage channel or closer to the edges of the channel or pool. The center of the drainage channel of a hot spring or geyser, being the hottest part, 73°C to 75°C limit, is either barren, or, more commonly, white with high-temperature bacteria. Many springs are edged with blue-green algae. The varied colors of the organisms produce a spectrum from which temperatures can be rather accurately estimated. In some springs the colors change from season to season, depending upon the average ambient temperature.

The higher temperature organisms are white, grading into light yellow, then orange, red, and finally dark green as the temperature decreases. The blankets of color result from vast assemblages of one cell bacteria and blue-green algae whose pigments within each individual cell cause the color. The most prevalent and important pigment is grass-green chlorophyll, which through photosynthesis absorbs and utilizes energy from the sun. All algae and some bacteria contain chlorophyll. A few algae also contain another pigment, phycocyanin, which absorbs sunlight but at a different wavelength than chlorophyll. The two together give some algae their blue-green color. Carotenoids, yellow, orange, and red pigments, are also present in all blue-green algae and a few bacteria. Sometimes they impart their color to those algae and bacteria deficient in chlorophyll. One type of bacteria, a carotenoid orange-red variety associated with alkaline springs, contains a special type of chlorophyll, indicating that at least some of its energy is derived from the sun.

Table 7-2. Temperatures of Biological Importance (Adapted from Brock, 1978)

	Temperature (°C)
Upper temperature for bacteria	>100
Water boils at sea level	100
Water boils at Yellowstone	93
Upper temperature limits for:	
Blue-green algae	70−73
Fungi	60−62
True algae	55−60
Protozoa	56
Crustaceans	49−50
Mosses	50
Hot spring flies	45−50
Higher plants	45
Vertebrates (frogs and fishes)	38

Essentially every neutral or alkaline hot spring in the world contains bacteria. They often flourish in boiling water, doubling their population in two hours. Unlike algae they do not form thick mat-like coverings but generally cling to sinter walls and pebbles. In the temperature range from 50°C to 60°C, in which both algae and bacteria prosper, a network formed by the long and narrow bacteria strands entrap the algae and prevent their being washed way. A mat as thick as 5 cm can form in channels even where constant flow occurs.

Animals also are abundant around hot springs. The most visible are the buffalo and elk at Yellowstone who invade the hot springs and geyser areas, eating the abundant algae and small plant growth while their normal feeding grounds are covered with heavy winter snows. They must also appreciate the warmth.

Less obvious but much more numerous are the small nonbiting *Ephydra bruesii* flies. They are found all over the world, mate and live in clusters in the cooler yet still warm waters of the hot spring drainage channels. Indigenous to alkaline springs, they lay masses of bright orange-pink eggs above the algae mat on projecting twigs and stones where, during winter, spring, and early summer when the algae mats are dark green, they are very conspicuous.

Parocoenia turbita, another indigenous type of fly, buries its white eggs in the mats where they are nearly invisible. The flies swarm on the warm mats during cool summer nights and early mornings; they are hard to find during the heat of the day. They circumvent their maximum survival temperature of 43°C while foraging for food in hotter waters by encasing themselves in an insulating air bubble. The larvae, which are extremely abundant, in some springs as high as $100,000/m^2$, develop best in the temperature range from 30°C to 35°C. An egg usually hatches within 24 hours, the larva crawling onto the mat to eat large quantities of it. A week later it transforms into a pupa, emerging as an adult fly a few days later. More than 500 flies will sometimes live in a square meter of area.

Ephydrid flies eat mostly algae and bacteria, but the *colichopodid* fly, indistinguishable from the *ephydrid* except it is larger and more long-legged, eats the eggs, larvae, and adult *ephydrid* flies. Dragon flies, spiders, tiger beetles, wasps, and killdeer all also feed on the *ephydrid* flies as well as several parasites, especially a small red mite.

Plant and animal life around acid springs is very different. It is neither as abundant nor does it exist at such high temperatures. The cellular structure of the algae, which includes a distinct nucleus, is more complex in acid pools than alkaline ones, enabling it to grow in extremely acid water. The upper temperature limit at which this algae will grow is 56°C, about 17°C lower than that for alkaline waters. Numerous bacteria also grow and survive at much higher temperatures.

7.4 Ground Noise and Seismicity

The description from Krug von Nidda of the ground noises of the Great Geysir, is quite dramatic. While standing about 60 paces from it:

> We heard a dull thunder-like noise under our feet, which soon became louder, and was changed to sounds resembling a series of shots, and following each other in rapid succession. The earth experienced a trembling movement; I hastened out of my tent and saw great masses of steam bursting from the interior of the geyser, and the water of the spring thrown out to a height of fifteen to twenty feet. This agitation of the geyser lasted scarcely a minute, and the usual perfect tranquility was then restored . . . This . . . was one of the smaller eruptions

occurring about every two hours. About a day later, after several smaller eruptions, all accompanied by similar sounds,

> . . . the roaring noise resounded from beneath . . . Twelve to fifteen formidable thundering reports followed, during which the ground was violently agitated by a vibrating movement. I hastened from the edge of the basin for it threatened to burst asunder under my feet . . .

Even though the Great Geysir no longer erupts, the ground around it continues to react to its inner workings. A typical record obtained by placing a seismometer on the lip of its basin is reproduced in Fig. 7-6. The principal motion is a slow undulatory heaving having a time constant of 10 to 15 s, with surface velocity of 1×10^{-6} cm/s. The motion could be produced by forces associated with the periodic upwelling of water in the geyser basin on the footings of the cone.

Records from seismometers placed on the ground close to a number of geysers indicated that each has its own characteristic ground motions generated by thermal and hydrologic activity within it.

Observations have been most extensive at Old Faithful. The first ground movements and noises occurring after its eruption presumably are associated with the movements of the subterranean water that rather quickly fills the geyser within about 20 minutes. These first seismic signals consist of a short series of long period movements lasting about a minute. Each individual pulse persists for about 10 s (Fig. 7-7). A similar series may be repeated shortly thereafter and then this type of signal ceases entirely. Occasionally the motion is accompanied by audible booming.

Then short, ¼ s long, high frequency, 20 to 50 Hz, pulses of seismic energy of the type shown in Fig. 7-8 begin to appear at a repetition rate of about 50/min. Some are generated by the collapse of steam bubbles that form in the warmer lower waters and rise into the cooler upper waters, but the majority are from the explosion of steam bubbles as they reach the surface. Visual observations show the quiet lazy venting of steam from the geyser opening which has gone on since

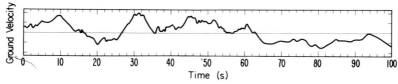

Figure 7-6. Ground velocity as a function of time at the rim of the geyserite cone of The Great Geysir. Record taken July 8, 1967.

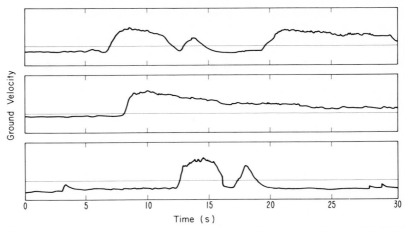

Figure 7-7. Low frequency ground motion on Old Faithful's cone. Record taken Jan. 1966.

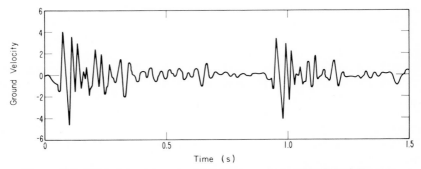

Figure 7-8. High frequency seismic pulses recorded at Old Faithful's cone.

the eruption, is now replaced by a continuing series of intermittent vigorous puffs of steam. Temperature measurements indicate that considerable circulation occurs after the high frequency seismic signals start, particularly at the lower levels.

A cumulative distribution of the number of the high frequency seismic signals as a function of time is shown for four successive eruptions in Fig. 7-9. The repetition rate increases with time from the initial 50/min to 90/min shortly before the eruption starts. The signals, as a rule, stop about 2 min short of the beginning of the eruption.

The intensity of the high frequency background noise, always present to some extent and characteristic of vigorously boiling hot springs, increases substantially 40 to 60 min into an interval (Fig. 7-10). This is essentially the same time that the temperature at the shallower depth levels starts to decrease, suggesting that the drop in temperature is the result of steady state boiling gradually working its way down from the top. The high frequency individual seismic pulses are now

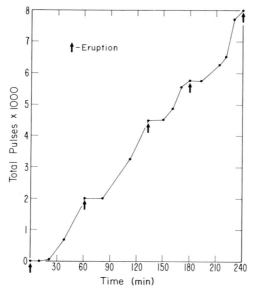

Figure 7-9. Frequency and temporal distribution of high frequency pulses. Activity typical of long and short intervals between eruptions of Old Faithful. Termination of each curve indicates time of occurrence of next eruption.

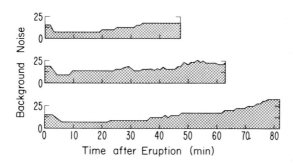

Figure 7-10. Background noise at Old Faithful. Arbitrary vertical scale.

coming from steam bubbles that are forming and bursting at greater depths, each pulse corresponding to the explosion of a bubble.

Other discrete distinguishable signals occur in addition to the high frequency pulses. Traces A and B, Fig. 7-11, exhibit quite weak one-sided signals recorded at slow and fast speeds, respectively. These occur at a rate of 1 or 2/min, even as often as 6/min, throughout the entire eruption cycle, observable even during periods when the high frequency pulses are not. The pulses are not truly one-sided. If they were, the cumulative effect of the motion would lead to substantial permanent earth movement which of course is not the case. Presumably after an initial impulsive displacement, likely caused by a small fracture suddenly open-

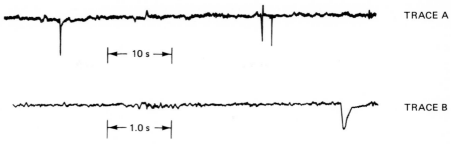

Figure 7-11. Weak one-sided pulses recorded at Old Faithful.

ing up, the ground recovers, but too slowly for the instrument to record the motion.

Reyholtshver Geyser, Iceland, although very much smaller, generates high frequency seismic pulses similar to those found at Old Faithful. It erupts from its small standpipe extremely regularly, at about 9 min intervals. Its seismic signature consists of a series of short, about ½ s duration, bursts of high frequency, 20 Hz pulses, spaced at intervals of about 3 s. These pulses start immediately after an eruption, stop abruptly after 2 min, only to begin again after the next eruption 7 min later. The performance of the geyser is remarkably regular as is evident in Fig. 7-12 where the number of high frequency bursts, always close to 40, is summed as a function of time through four successive eruptions.

Riverside Geyser in the Upper Geyser Basin, is the most regular geyser in Yellowstone. Its approximate 6 hr interval between eruptions never varies by more than a half hour either way. The geyser erupts as a 25 m high, powerful, inclined jet from a small hole in a throne-like sinter pad sitting on the east bank of the Firehole River about 2 km downstream from Old Faithful (Fig. 3-4). The jet plays to its maximum height for about 5 min, and then dies down gradually, stopping altogether in another 15 min. Its seismic signature, as might be expected, is similar to that of Old Faithful since they are both columnar geysers. It differs mainly in two aspects: the pulses contain higher frequencies, usually 50 to 60 Hz, and occasionally 80 to 90 Hz; and the amplitudes of the pulses vary in a characteristic way by over a factor of 10 throughout the interval between eruptions. At times, each high frequency pulses consists of two to four oscillations, often occurring in pairs spaced a fraction to 2 to 3 s apart. At other times, a pulse consists typically of 15 to 20 distinct pulses in a 10 to 20 s time span with short periods of quiet in between. One-sided, sawtooth-shaped seismic signals of the type generated by a sharp blow or release of strain are also present several minutes before eruption. These are probably caused by small fractures suddenly developing close to the geyser opening. Unlike the relatively weak one-sided pulses at Old Faithful, these are strong, the same magnitude as the high frequency signals. The smallest signals at Riverside are found before the onset of overflowing water which occurs about 1.5 hr before an eruption. Shortly after that, they increase in magnitude by more than a factor of 10 remaining high with

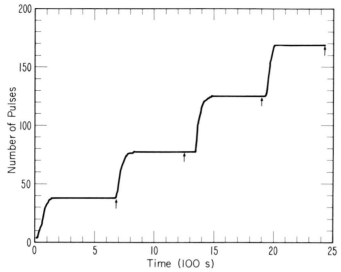

Figure 7-12. Cumulative time distributions of high frequency pulse generated in the ground at Reykholtshver Geyser. Arrow indicates eruption. Record taken July 2, 1967.

considerable variation, until the eruption occurs. They still continue after the eruption but at an intermediate level until about 3 hr before the next eruption when their intensity drops by a factor or two for an hour or so until the water starts overflowing and it again increases manyfold.

The frequency content and the ground velocities of the seismic pulses generated by the nine Yellowstone geysers that have been monitored are essentially the same, usually ranging from 10 to 60 Hz and 0.10 to 5.0×10^{-3} cm/s, respectively (Table 7-3).

The rate of occurrence of the pulses varies greatly from geyser to geyser and during the interval between the eruption of any particular geyser. In most cases, the seismic behavior can be correlated with some specific observable geyser action. Lone Star Geyser is an exception. Boiling and eruptions produce the signals at Bead, Jewel, Plume, and Seismic; one-sided impulsive-type seismic pulses, recorded at Old Faithful, Riverside, Castle, Plume, and White Dome are indicative of ground heaving and breakage, followed by abrupt falling, probably related to mass movements of water. The most informative signals are obtained at Old Faithful where there is a strong correlation between the time of onset of seismic activity and the length of time to the next eruption.

The seismicity of Strokkur is characterized by a series of four or five short, 1 to 2 s bursts of oscillations (Fig. 7-13) which always follow an eruption. The first two or three of the bursts in the series can be heard at the surface. The first burst occurs about 9 s after the eruption with the rest following along at 2 or 3 s intervals. They seem to be associated with the sudden refilling of the subterranean reservoir, most likely by the plunging downward of near-surface water

Table 7-3. Frequency Content and Magnitudes of Seismic Pulses Generated by Geysers (After Nicholls and Rinehart, 1967)

Geyser	Predominant Frequency (Hz)	Magnitude of Surface Velocity (10^{-3} cm/s)
Old Faithful	30−50	2.5
Riverside	50−60	1.8
Castle	20−40	0.1
Bead	30−40	2.0
Plume	50−60	2.5
Jewel	6−10	1.3
White Dome	15	2.5
Seismic	50−60	5.0
Lone Star	30−40	1.3

Time After Eruption (s)

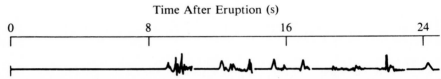

Figure 7-13. Ground velocity as a function of time at the rim of Strokkur Geyser. Record taken July 7, 1967.

which would produce a water hammer effect. The burst signals are one-sided, the main ground motion being upward as if it were being hit from below. An examination of the records taken at high recording speeds show that successive bursts are not identical; instead, their shapes change progressively. The first burst consists of three distinct, closely spaced, roughly sawtooth-shaped pulses on which are superimposed oscillations of about 20 Hz. Each successive pulse is probably associated with a slap of water against the wall of a reservoir. They are progressively more extended with the individual pulses being more spread out, suggesting a slowing down of the sloshing of water from reservoir wall to reservoir wall. The similarity of pattern from eruption to eruption is remarkable.

Similarly appearing and functioning geysers have different seismic signatures. South Geyser, Beowawe, and Uxahver in northern Iceland, are such examples. South Geyser has an opening 30 cm in diameter which opens into a shallow 4 m diameter, irregularly shaped catch basin. During the eruptions that occur about every 4 min, water plays to a height of about 2 m for approximately 1 min, with a great deal of water overflowing. Between eruptions, the water in the pool stays at a more or less constant level with occasional bubbles of steam and superheated water welling up and exploding at the surface. Uxahver has about the same size pool and erupts explosively every 2 or 3 min, throwing up a 2 m high plume, lasting about 15 s. Sometimes it is difficult to distinguish between a violent

boiling phase and an actual eruption. In both geysers, as the eruption time approaches, the intensity of the background noise arising from the boiling within the pool builds up and reaches a maximum just before an eruption. There the similarity ceases.

South Geyser's eruption is followed by an abruptly initiated, sustained upward motion of the ground lasting for about 1 s, which stops suddenly only to begin again, but more slowly, the total upward motion lasting for about 45 s. The upward displacement, which is followed by partial recovery, is probably generated by downward rushing water refilling emptied subterranean cavities with the first sudden motion caused by the initial abrupt slap of returning water against the roof of the cavity.

Uxahver's behavior is just opposite. A few second after an eruption, the ground begins to subside at a velocity of about 2×10^{-4} cm/s as contrasted with South's rise of 8×10^{-6} cm/s. The subsidence continues at this constant rate for a time ranging from 1 to 20 s, but averaging about 8 s. After certain eruptions, the subsidence stops briefly, sometimes more than once. When subsidence continues for a relatively long time, more than about 8 s, the stoppage is more prolonged, occasionally as much as 5 s. In all cases, the ground surface finally slowly stops moving downward and begins moving upward, coming to rest after several minutes. The subsidence must be caused by collapsing of the reservoir emptied during the eruption; and the recovery, by refilling of the reservoir.

Temporal Changes in Geyser Activity and Their Causes

8.1 General Nature of the Changes

Some geysers erupt quite regularly, others sporadically; some are active for a while, then become inactive, only to rejuvenate at a later time; new geysers come into being while others fade away. There is continual change in geyser activity, partly periodic, partly secular, and partly sporadic. But there are also remarkable threads of constancy.

Certainly continental drift, mass transport of material by glaciers, rivers, erosion, and mineral deposition are important agents in initiating and regulating changes in geyser activity. The time scales associated with these processes range from millions of years to a few days or hours. But many of the changes in geyser activity are caused by other mechanical forces active within the earth. These keep the underlying gravels and fractured rock in constant motion, causing the gravels to slump or creep, changing the porosity of the rocks, and opening and closing fissures. Such activities affect the flows of gases and hot water. The most effective mechanical agents are:

- Barometric pressure, which pushes against the surface of the earth;
- Buildup and release of stresses associated with earthquake activity and intermittent but noncatastrophic earth displacements; and
- Variable gravitational pull of the sun and the moon on the earth.

The earth is so large, massive, and mechanically sluggish that the effects of changes in forces can be delayed months and years. Moreover, the specific mode and degree of the reaction of any particular geyser will be determined by its geologic setting and plumbing system.

A common impression of a geyser, even among earth scientists, is that it functions on a regular predictable schedule: Old Faithful. This is certainly not the

case. The time intervals between eruptions of Old Faithful vary by a factor of three to four, from between 30 to 40 min and 100 to 120 min. The temperatures and flow rates of most hot springs and geysers are continually changing, some only very slowly over long periods of time, others abruptly from time to time. Many exhibit regular and cyclic fluctuations. As a result, the frequency of eruptions, the duration of play, and the heights of the eruptions will all change. In the Haukadalur Valley in Iceland, extensive layers of old disintegrating and often buried siliceous sinter cover large areas where springs no longer exist but where they must have been in earlier times. Further down the rift valley, new springs are forming fresh sinter. In Yellowstone, many new geysers have started up since the first organized expeditions 100 years ago. Several have become extinct, at least for the time being, either going dry because its waters found a new outlet or becoming a nonerupting hot spring by losing part of its heat supply.

Strokkur's eruption pattern has become quite different over a 130 year period. An eyewitness describes an 1837 eruption:

> A thick column of smoke ascended suddenly to the clouds . . . inclined jets of water forming paths for themselves through the smoke; some attained a height of 100 feet. Large stones which we had previously thrown into the spring were projected . . . Afterwards the column . . . consisted only of steam, which left the mouth with a whistling and hissing noise, and rose with incredible velocity towards the clouds. The activity of the spring continued in this manner uninterrupted for three quarters of an hour, when tranquillity was restored . . . According to the natives . . . eruptions occur only once in the two or three days.

By 1967, it had calmed down a great deal, erupting much more frequently:

> Every 10−15 minutes, the geyser erupts. When an extra large or extraordinary hot mass of superheated water reaches the surface the water changes suddenly to steam, and the explosion hurls a small quantity of water to a height of 20 meters or more. The eruption is short lived, a matter of only a few seconds.

In New Zealand, many changes have been caused by man himself. The Wairakei geothermal electric power development, by extensive extraction of hot fluids, has completely dried up an entire geyser basin. The formation of a lake that is part of a hydroelectric power project has flooded practically all of the geysers in the Orakei-Korako area.

A geyser might quench its own activity by the deposition of its sinter either by clogging up the channels supplying heat and water to it or by building its cone and other surface appurtenances up to such heights that the available heat will not bring the water to a boil against the increased hydrostatic pressure. The former probably happens fairly frequently. At the now inactive Te Waro Geyser, the deposition of silica in porous subterranean rock has decreased its permeability in this way. Quenching by cone building happens more rarely; Castle, Lone Star, and White Dome Geysers, all of which are very active, issue from massive, high cones which at some future time will seal themselves off.

The eastern edge of the Porcelain Basin, Norris, has undergone more thermal

change during the last 100 years than any other area in Yellowstone. Valentine Geyser played for the first time on Valentine's Day in 1902 from a deep, steaming hole. The Harding Geyser, named for the then President, first erupted in 1923 and disappeared soon after the President's visit to the Park. Monarch Geyser, which by 1935 had become a warm stagnant pool of muddy water, at one time was one of the principal geysers in the Park. Ledge Geyser, sometimes a spectacular geyser issuing at an angle from the side of a cliff, was at one time known as Muddy Geyser, remarkable for the thick mud it contained which was periodically thrown to a considerable height. The changes in Steamboat's activity have been phenomenal.

The immediate cause of the fluctuations in thermal activity, appearance and disappearance of mud in springs, and sudden jetting of superheated steam at Norris is due to movements of the unconsolidated gravels and clays in which the geysers are located. The shifting of these would close some subterranean channels and open up others, impeding and facilitating the movement of hot water and steam through the complex interlocking network of passages by which heat and water reach the springs. Changes in thermal activity do not imply a change in the constancy of heat supplied to a general area but only local rerouting of that heat and perhaps the water supply.

The activity of six geysering wells drilled in Japan in 1939 changed substantially over a 15 year period. Progressive changes in one 18.5 m deep, 10.5 cm diameter well were as follows:

1940: Two types of eruptions were occurring, a fairly irregular one playing to a height of 3 m for a time ranging from 3 to 85 s every 3 to 24 min. Because of its relatively short interval between eruptions, this type is referred to as a "Kobuki". The second type was an "Oyabuki", a less frequent but larger eruption, playing twice a day for 14 min to a height of 15 m.

1941: The Oyabuki was no longer playing; the Kobuki was playing to a height of 2 m for 10 s only every 5 min.

1942: The Kobuki was now playing only every 15 min, but for a longer period of time, 23 s.

1947: All eruptions ceased but just prior to this, Kobuki was playing for 42 s every 54 min.

1951: Oyabuki eruptions were resumed at 100 min intervals, occasionally playing as long as 30 min.

The causes of the changes have not been reported.

8.2 Bimodal Eruption Patterns

A characteristic of many geysers is their two or more distinctive sets of intervals between eruptions, a group of relatively short intervals and a group of relatively long ones such as the Kobuki and Oyabuki eruptions exhibited by some Japanese geysers. In some other geysers, long and short intervals follow in perfect sequence as at Narcissus, and at others, such as Old Faithful, they are more or less

random. Bimodality in the distribution between eruptions was recognized at least 100 years ago at Yellowstone. The 1879 annual report of the Superintendent reads:

> . . . (Steamboat) seems this year to have settled down to business as a very powerful flowing geyser, having in common with many others, a double period of eruption, one some 30 feet high about each half hour, and another of nearly 100 feet and long continued, each 6 or 7 days, and is doubtless still changing.

It appears more recently that Steamboat in fact exhibits three types of eruptions: the most frequent occurring at 2 to 3 min intervals jetting water 5 to 8 m; a more forceful one every 4 to 5 min, shooting water to heights of 12 to 15 m; and finally its infrequent tremendously powerful eruption. Similarly the Great Geysir, when active, erupted to a height of about 6 m every 2 hr and to a height of 25 to 30 m every 24 to 30 hr.

The eruptions of Narcissus follow a regular sequence of alternating 2 and 4 hr intervals. Measurements indicate that the rate of influx on the ground water filling its basin is essentially constant, but the eruption terminating a short interval occurs before the basin is filled, whereas water runs over the brim for half an hour or so before the 4 hr interval is concluded. The marked difference in behavior of successive intervals could possibly be the result of retention of heat in the porous geyserite basin walls during the storage of water preceding the long intervals which tends to shorten the heating time required during the short interval. It seems more likely however, that it is related to the emptying and filling of underground reservoirs.

Viewed over a 100 year span, the average interval between eruptions of Old Faithful has remained remarkably constant, a little over 65 min. However, in some years they are bimodally distributed with one mode averaging 50 min and the other 70 min, the latter being about twice as prevalent as the former. This is evident in Fig. 8-1 which shows the distributions for several selected years between 1878 and 1977. The shape of the frequency distribution curve has changed radically. From 1870 up to about 1920, although only a few records exist, there is no suggestion of excessively short or long intervals. The first indication appears in 1922 when the Superintendent's report for 1921-22 mentions extreme intervals of 36 and 82 min. The Yellowstone Nature Notes for September 1922 states that "the intervals of Old Faithful are from 60 to 70 minutes, but lately this geyser has shown some abnormally short periods." In the mid 1930s, long and short intervals are mentioned again in the reports.

The first definite indication of a biomodal distribution appears in 1941 and by 1943 it was quite pronounced, its respective modes being 48 and 69 min. Such a bimodal distribution persisted through the 1940s, but by early 1950 a lack of short intervals had erased the lower mode. 1953 shows no lower peak. During the late 1950s it began to return and by 1959 the lower peak had become very prominent. The pronounced biomodal character of the curve was maintained until the mid 1970s when it began to disappear so that by 1977, a 75 min eruption was 3.5 times as likely to occur as a 50 min one.

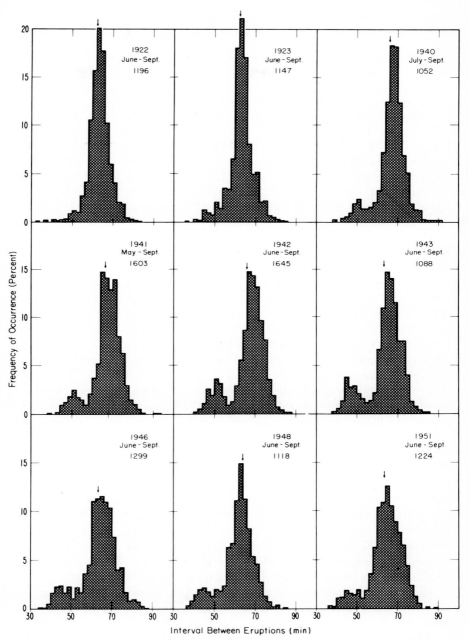

Figure 8-1. Frequency distribution by year for intervals between eruptions of Old Faithful. Dates and numbers of intervals indicated.

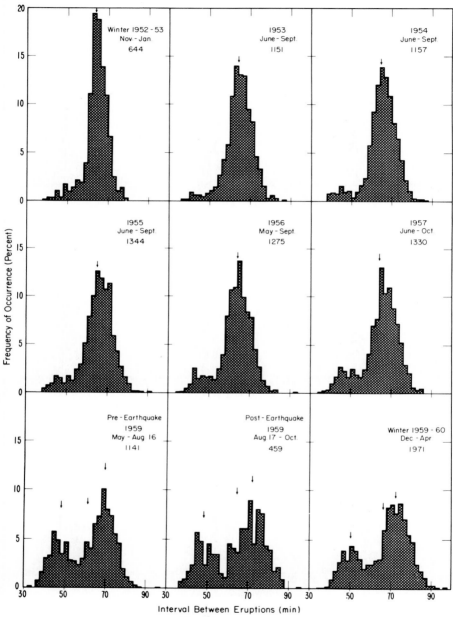

Figure 8-1. *(continued)*

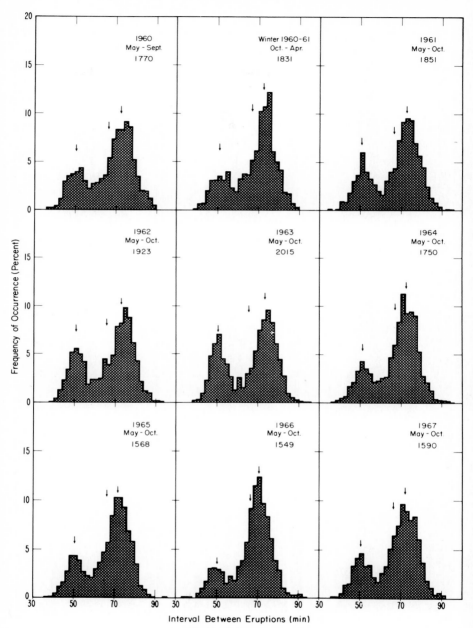

Figure 8-1. *(continued)*

There is no fixed sequential pattern of interval lengths. On certain days long intervals alternate with short intervals in almost perfect rhythm; on other days, all intervals will be about the same moderate length, 60 to 65 min; and on still other days, the pattern is quite erratic, a short interval being followed by one or more long intervals with moderate length intervals mixed into the sequence. Sequences typical of these patterns are shown in Fig. 8-2. Only very rarely do short intervals follow one another. Unfortunately continuous readings for several days in a row are available in only a few cases as they are generally taken only during the daylight hours.

It follows that the length of the active play of water from Old Faithful, varying from 2 to 5 min, will also be bimodally distributed (Fig. 8-3). To a very close approximation, the length of the interval that follows an eruption varies linearly with the length of play preceding it, a long play heralding a long interval and a short play, a short interval. Since short intervals almost never follow one another, a short interval will generally be followed by a long play. On the other hand, a long interval may be followed by either a short interval or a long interval as well as either a short play or a long play.

Seismic observations of ground motion indicate the length of the interval between eruptions. Figure 8-4 is a plot of the time after an eruption at which the high frequency seismic pulses (Fig. 7-8) start as a function of the ensuing interval. When the interval is to be short, the seismic pulses begin almost immediately after the previous eruption, whereas if the interval is to be long, onset of seismic pulses will be delayed for 15 to 20 min.

It is instructive to speculate on the nature of the plumbing responsible for Old Faithful's pattern of bimodal behavior. The model shown in Fig. 8-5 could account for it. The reservoir is considered to be a single, roughly circular cylinder of sinter which penetrates to a depth of a few hundred meters. Warm local ground water flows into the tube near its top, perhaps between 10 and 20 m down, at a more or less constant rate. There is also a channel, probably draining into the nearby Firehole River, through which the geyser tube, when full, can

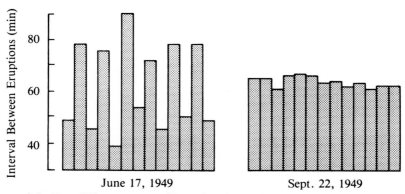

Figure 8-2. Two different patterns illustrating day to day irregularity in Old Faithful's sequence of long and short intervals between eruptions.

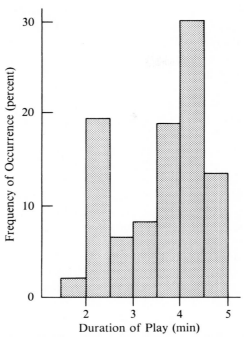

Figure 8-3. Frequency distribution of lengths of play of Old Faithful during 1965 summer season.

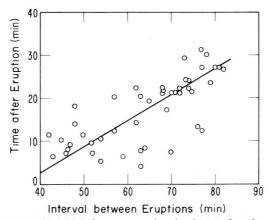

Figure 8-4. Time at which high frequency pulses begin as a function of interval between eruptions of Old Faithful.

discharge water at the same rate that it flows in. Temperature measurements indicate that this channel is about 10 m below ground surface. The water in the tube is heated by injection at considerable depth of hot water.

A long play simply empties more of the tube than a short play, requiring a longer time for the water rising in the tube after the eruption to develop sufficient

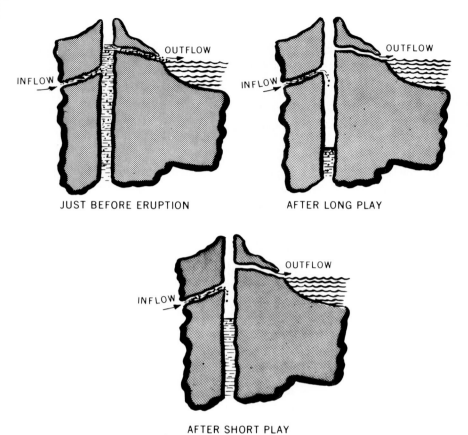

JUST BEFORE ERUPTION   AFTER LONG PLAY

AFTER SHORT PLAY

Figure 8-5. Inferred plumbing of Old Faithful to account for bimodality of intervals between eruptions.

hydrostatic head to suppress normal boiling, a necessary prelude to the initiation of seismic pulses which are generated by intermittent steam bubble formation and collapse.

Differences in temperature regimes exist once the tube is filled and overflowing. Of great importance, especially with respect to functional bimodality, is the fact that when the tube is full following a long play, the amount of top cooler water is much greater, perhaps twice as much as it is after a short play. Thus, after a long play much more time will be required to develop the critical temperature distribution needed to precipitate an eruption than would be required after a short play.

These observations are consistent with the model. Following a short play, the amount of heat stored in the geyser reservoir is greater than after it has been emptied by a long play; hence, assuming that rate of heat input to the reservoir is essentially constant, the time to the next eruption, after being only partially emptied by a short play, should be relatively short. Vice versa, the time should

be long after a long play. And medium length intervals should follow medium plays.

Also consistent with this model is the observation that the eruption jet structure is essentially the same for both short and long plays. Water is ejected intermittently in rapid staccato, propelled by a series of steam explosions at velocities ranging from 10 to 30 m/s in about 8 to 10 thrusts/min. Initially the action is mild; it builds up over a 15 to 20 s period to a height of about 25 m after which it tapers off. The jet during a long play is distinguished primarily by its persisting low level activity following the main buildup and decay. The similarity in initial behavior seems to indicate that conditions near the top of the geyser tube, the region where the eruption is initiated, are essentially identical at the onset of both long and short eruptions.

8.3 Climatic, Seasonal, and Barometric Effects

As geyser action depends on water, and indeed they are only found in well watered areas, it is logical to assume that weather, especially variations in the amount of rain and snow, would strongly affect a geyser's actions. Actually this is not the case except in a few instances and except for some near-surface features, for example the mud pots which become quiescent during a dry season. One of the geysers whose activity is affected is Waikite at Whakarewarewa. Its vent is the most conspicuous sinter mound in the area, and the water level within it fluctuates markedly depending on rainfall. When the level is high, the geyser erupts 15 m high pulsating jets. During times of low rainfall, the water level drops below the 7 m deep geyser tube and it is transformed into an intermittent fumarole.

Atmospheric temperature variations do not seem to affect geyser performance except in a few isolated cases. A strong, cold wind blowing across the broad surface expanse of a fountain geyser, for example, may delay its eruption. Observations made of some 50 eruptions indicate that this will happen at Daisy Geyser where the intervals between eruptions ranged from 153 min to 90 min, averaging 111 min. The longest intervals occurred when strong north-south winds were blowing over the geyser.

The latitude and altitude of Yellowstone subject the geysers there to one of the most rigorous climatological conditions in the United States. Mean temperature ranges about 20°C from the summer six months to the winter months. These temperature differences do not seem to have any significant effect on the principal geysers, including Daisy which is affected only by local winds. Summer and winter, day in and day out, the main water supply and the heat supply are insulated from atmospheric temperature changes by a sufficient cover of soil and rock so that the temperatures of both remain essentially constant.

Variations in barometric pressure, though seldom more than about one percent, do have a noticeable effect on geyser activity. It is a well-established fact that the temperatures of many hot springs increase a few degrees under reduced

atmospheric pressure. Before the Atami Geyser ceased erupting altogether, low atmospheric pressure lengthened the intervals between eruptions and high pressure shortened them. The most likely explanation is that the rock or soil mass expands somewhat under the reduced pressure, widening veins and fissures and opening up pores to allow increased flow of hot water and gases.

Both Old Faithful and Riverside Geysers respond to barometric pressure variations but the response is quite erratic. At times it is impossible to establish any correlations that are convincing, possibly because the buildup of tectonic stresses in the geothermal area adversely affect their ability to respond to changes in barometric pressure.

One of the most carefully studied geysers vis-á-vis effects of barometric pressure is the geysering drilled well, Old Faithful of California, located at Calistoga. Here the effect of the annual cyclic variation in barometric pressure is very prominent. Everywhere in the world, the average barometric pressure goes through an annual cycle, the times of occurrences of the maxima and minima varying from place to place. At Calistoga, the spread between maximum and

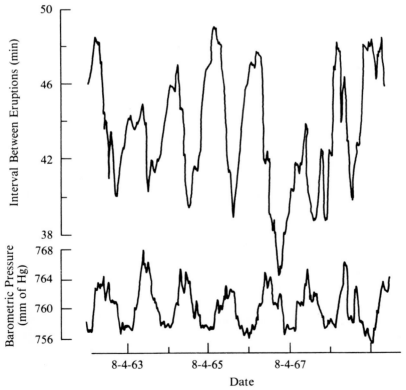

Figure 8-6. Upper curve: interval between eruptions (averaged over 30 days) as a function of date for Old Faithful of California Geyser. Lower curve: average barometric pressure during the same period.

minimum ranges from 6 to 20 mm of Hg, with the maximum occurring during the three month period from November through January. Surprisingly this is coincident with the rainy season when the average barometric pressure might be expected to be low.

The 30 day running average of intervals between eruptions is plotted as a function of time over an approximate seven year period in Fig. 8-6. Juxtaposed is the curve of barometric pressures. The high barometric pressure in winter shortened the interval to 40 min and low summertime barometric pressure lengthened it to 49 min. The negative correlation between barometric pressure and the intervals between eruptions would be made much stronger by an approximate 70 day shift in the curves which improves the computed correlation coefficient to 0.9. This simply indicates that the sediments filling the basin are slow in responding to the action of an applied mechanical force and further that the total displacement is more important in bringing about change in geyser action than stress amplitude. Old Faithful of California sits in a highly water-saturated bed of fine gravel which acts like a sponge with the high and low pressures alternately allowing the hot water to flow in and to be squeezed out.

8.4 Earthquake Effects

Since geysers generally are located in tectonically active areas, earthquake country, they are subjected to the stress changes preceding, occurring at the time of, and following earthquake activity. The general pattern of change is related to variations in tectonic forces which affect the rate of flow of fluids, both gases and liquids, within the fractured and porous rock masses by modifying the dimensions of flow channels and pores. Large earthquakes release as much as 10^{25} ergs of energy that has gradually been stored up in the rock mass as it becomes more and more strained. This storage takes place over varying periods of time, in general the length of time varying linearly with the logarithm of the magnitude of the ensuing earthquake (Fig. 8-7). For smaller 2 to 3 magnitude earthquakes only a few days are involved; whereas for larger ones stresses build up for two to three years ahead of time.

Geyser areas are laced with faults. The quake is precipitated when the stress associated with it exceeds in some small region the frictional resistance holding the two sides of one of these faults in place. The suddenly released energy generates a radiating seismic disturbance. The release of energy is not instantaneous and usually a single earthquake does not completely relieve the strain. It may be preceded by numerous small foreshocks, and aftershocks may continue for several months afterward.

That geyser performance is responsive to earthquake activity has been observed in Iceland for centuries. The event having the most profound effect in Yellowstone since its discovery was the close-by, large magnitude 7.1 Hebgen Lake earthquake, August 17, 1959. The epicenter was only about 48 km northwest of Upper Geyser Basin. No earthquake of comparable magnitude has oc-

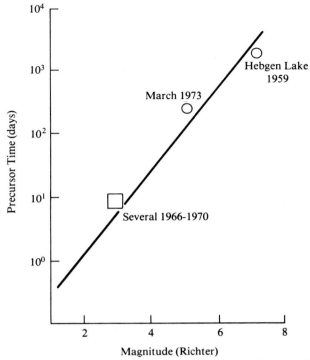

Figure 8-7. Log of time at which precursor activity begins as a function of magnitude of ensuing earthquake. Circles and squares indicate time at which geyser activity begins to change.

curred near the geyser basin in historic times. It precipitated more hydrothermal fluctuations than had occurred during the previous 90 years.

These dramatic changes were at once and carefully studied. All active geysers erupted immediately following the earthquake. The average temperatures of the waters of 73 geysers increased by 1.3°C and that of 89 pools and springs by 4.8°C. Many long dormant geysers began erupting; a few new geysers sprang up; the frequency of eruption of already active geysers increased. In general, the efflux of heat became larger, probably as the result of opening of fissures and fractures, and remained abnormally high for several years. New hot ground developed, especially along the Firehole River. Surface cracks and breaks appeared in all three basins, but was especially evident in the Lower Geyser Basin. Some of the fractures were new and others were ones previously sealed by mineral precipitation. Clear pools became muddy; fumarolic activity developed along the cracks, which in one instance evolved into Seismic Geyser, a geyser of substantial size. Its growth pattern is typical of many other older geysers. It began as a crack opened up by the earthquake, became a fumarole, and then over a period of several years developed into a large geyser, only to decline as its energy was sapped by new cracks opening up.

The activity of Old Faithful was closely monitored. Surprisingly, for the first few days, it did not change appreciably. Later it behaved just opposite to most of the other geysers: it became less active with its intervals between eruptions increasing rapidly, most likely because of a decrease in the rate at which heat was supplied to it. For the early preearthquake period of the summer of 1959, its average interval was 61.8 min; by September 1, about two weeks after the quake it had increased slightly to 62.1 min; and it continued to increase, becoming more than 64 min by January 1, 1960. This behavior is shown in Fig. 8-8, a graph of the seasonal mean values of the time intervals between eruptions for the period from 1954 through 1970. The value of the mean time remained relatively steady from 1954 through 1957. After 1957, it began to decrease rapidly and during the succeeding three year period, from 1957 to August 1959, went down from 64.0 min to 61.3 min. After the night of the earthquake, the interval began increasing rapidly, so that by the summer of 1960, it had reached 66.5 min. The average interval remained high, even increasing slightly through 1961, at which time it then began to decrease in a regular manner, so that by the end of the summer of 1963, it had dropped from 67.0 min to 65.5 min. Sometime between the summers of 1963 and 1964, it increased noticeably to 67.1 min, very likely as a result of the distant but very large 8.4 magnitude Alaskan earthquake of March 1964. That such a distant geologic event would influence Old Faithful's behavior is not too farfetched in view of the fact that the quake affected water levels in wells throughout the United States and produced a small permanent displacement in

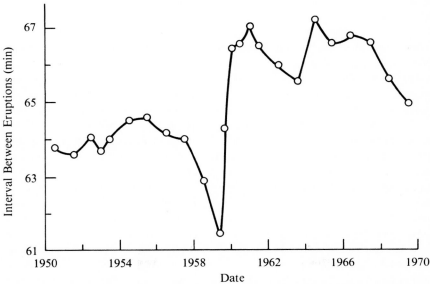

Figure 8-8. Yearly (summer season) average of intervals between eruptions of Old Faithful, 1950–1970, preceding and following Hebgen Lake earthquake of August 17, 1959.

distant Hawaii. Earthquakes seem to have little influence on the bimodal pattern of length of intervals characteristic of Old Faithful.

The post-earthquake effects are not surprising but what is, is the regular progressive increase in activity or decrease in interval, especially evident before the Hebgen Lake earthquake, which began two to three years previous to the quake. This suggests that the regional stresses within the earth began progressively changing two to three years before the quake and that they produced highly localized but significant alterations in the fissures and other openings supplying the hot steam and water that heat and activate the geyser.

However, when the data covering the time just previous to the earthquake are examined in detail month by month no obvious precursory behavior heralding the quake is found.

The immediate changes associated with the aftershocks were as large as those associated with the much more intense main shock. For about two days preceding a shock, the interval between eruptions steadily decreased, the decrease totaling 3 or 4 min. It then increased abruptly at the time of, or closely following the shock, finally ending at a level somewhat higher than that at which it started.

The leisurely reaction of Old Faithful, a matter of days, suggests that variations in permanent strain propagate at a relatively slow rate, a few kilometers per day so that it takes several days for them to be felt by a geyser located tens of kilometers away. Such behavior is consistent with the general notions of tectonic behavior patterns.

Another similar pattern of change was observed in Old Faithful during the last half of 1972 and the first part of 1973 (Fig. 8-9). A shortening in intervals between eruptions began in August, the monthly average decreasing from 66.4 min and reaching a 63.2 min minimum by February 1973. A series of earthquakes, 20 to 30 km from Old Faithful, near Yellowstone Lake, began with a magnitude 3.4 quake on 25 March. All told, nine quakes of measurable magnitude ranging from 3.4 to 5.0 occurred during the period from 25 to 31 March. The series was terminated by a 4.4 quake on 21 April. By June of the same year, the length of interval had risen to 67.8 min, a little above normal.

During the past 100 years, Old Faithful has responded to every other major local earthquake. Long term variations in intervals between eruptions are plotted in Fig. 8-10 for the period 1870 to 1970. The early data although fragmentary and sparse are most likely quite reliable. The time of occurrence of each major earthquake occurring within 100 km of Old Faithful is indicated by an arrow placed above the curve and its reported Mercali intensity by an appropriate Roman numeral. The pattern is clear and consistent. From two to four years before every major earthquake, the interval begins to decrease; it reaches a minimum near the time of the quake and then rises for a few years. By and large the curve is fairly flat, averaging about 65 min over the years, except in 1918 and 1920 when it went to a very high value, in excess of 75 min. It is tending to follow this same course in the late 1970s. The cause for the high values has not been established but it is probably associated with extremely low earth tidal forces that existed about 50 to 60 years ago and are again beginning to develop.

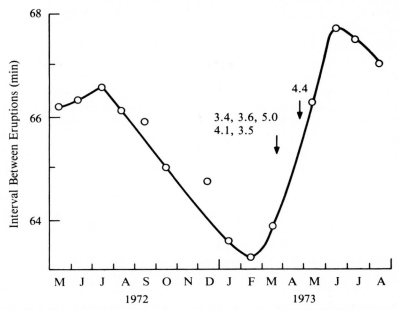

Figure 8-9. Fortnightly averages of intervals between eruptions of Old Faithful, summer, fall, and winter of 1972 to spring of 1973, spanning earthquake activity of March and April, 1973.

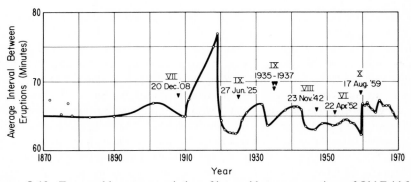

Figure 8-10. Temporal long term variation of interval between eruptions of Old Faithful. Roman numerals refer to Mercali intensities of earthquakes that occurred on dates indicated.

Riverside Geyser reacts somewhat differently. No detailed data are available on its performance shortly before and after the Hebgen Lake earthquake; however data are available for several smaller earthquakes occurring during the period from 1966 to 1969. The pattern exhibited by the curve in Fig. 8-11 is characteristic. A few days before the earthquake, the interval increases; just after, it goes through a minimum, which is then followed a few days later by a lesser minimum.

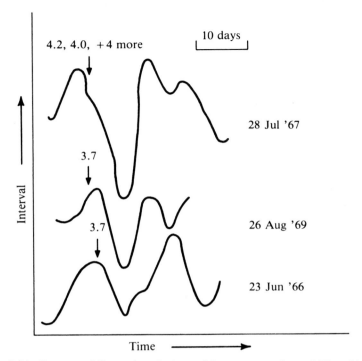

Figure 8-11. Patterns of fluctuations in interval between eruptions of Riverside Geyser shortly before and after occurrence of local earthquakes. Numbers refer to Richter magnitude.

The pattern of change in geyser activity preceding an earthquake is quite similar to that observed in other forms of geophysical activity: elastic wave velocity, gas emission, water level in wells, and electrical resistivity. The straight line in Fig. 8-7, showing the relationship between earthquake magnitude and time of onset of precursor changes, is derived from observations of such phenomena. The Yellowstone geyser data fit the curve well.

It is not uncommon for stress release to take place noncatastrophically in a matter of hours, days, or months. This sometimes produces observable surface differential displacements. That such movements are constantly at play is reflected in variations in geyser performance.

8.5 Earth Tidal Effects

The same forces arising from the gravitational attraction between the earth, sun, and moon that cause the ocean tides also generate earth tides, deformation of the rocks and soils of the earth itself. The not-so-solid ground beneath our feet is rising and falling about 20 cm every day. Thus the supply of heat and water to the geysers is being affected by alternate dilating and squeezing of the porous and

fractured surface of the earth. Dilation occurs under strong, high tidal forces that stretch the earth; squeezing action occurs under low, weak tidal forces. The regular twice-daily, daily, fortnightly, and semiannual tides are familiar ones, but there are also longer cycles caused by more complicated interrelated movements of the earth, sun, and moon. This constant mechanical agitation of the earth frequently influences and regulates important geologic processes and geophysical phenomena such as the flow of gas from rock masses, movements along stressed faults, and the activity of volcanoes, earthquakes, and geysers.

The figure of the earth, the masses of the sun and the moon, and the relative motions of the three bodies are well known, and it is possible to calculate in a straightforward manner the stress impressed on all points of the earth as a function of time. Newton's Universal Law of Gravitation:

$$F = G \, M_1 \, M_2/R^2$$

is used for this purpose where F is the force of attraction between two bodies of respective masses M_1 and M_2 whose centers of mass are separated by a distance R; G is the universal gravitational constant, an experimentally determined quantity, $6.67 \times 10^{-8} g^{-1} cm^3 s^{-2}$.

It is usually most convenient to consider the vertical and the horizontal components of the tidal forces separately. The vertical component, Δg_r, measures the amount that gravity is changed due to the presence of the sun and the moon; the horizontal Δg_ψ, influences the deflection of the vertical. Assuming a rigid earth, Δg_r is given by

$$\Delta g_r = -(a^3 g/M_E)\left\{\left[M_S(3 \, \cos^2\psi_S - 1)/R_S^3\right] + \left[M_M(3 \, \cos^2\psi_M - 1)/R_M^3\right]\right\}$$

and the total deflection of the vertical α is given by

$$\alpha = \tan^{-1}(\Delta g_{\psi_S}/g) + \tan^{-1}(\Delta g_{\psi_M}/g)$$

where ψ_S is the angle between a line connecting the centers of mass of the earth and the sun and the radius vector from the center of mass of the earth and a point on the earth at which the tidal force is to be calculated; ψ_M is the corresponding angle for the earth-moon pair; R_S and R_M are, respectively, the distances between the centers of masses of the sun and the earth, and the moon and the earth; and M_S and M_M, respective masses of sun and moon.

The maximum perturbing effect on gravity of the sun, about one half that of the moon, is given exactly by

$$\Delta g_S/\Delta g_M = (M_S/M_M) \, (R_M/R_S)^3 = 0.46.$$

The mass of the moon is relatively very much smaller than that of the sun but it is closer, with $R_S = 1.50 \times 10^8$ km and $R_M = 3.84 \times 10^5$ km.

Due to the rotation of the earth and the relative motion of the sun, moon, and earth, both distance, R, and angle, ψ, are functions of time and cause the tidal forces to vary continuously.

The axis and period of rotation of the earth and the equations governing movements of the sun and moon with respect to the earth are well established so

that the R's and ψ's and hence Δg_r and α can be readily calculated using electronic computers. The variations in both Δg_r and α are nearly periodic, having the following approximate periods: semidiurnal, diurnal, fortnightly, semiannual, 4.4 years, 8.8 years, 18.6 years, some multiples, especially 3, of 18.6 years, and 20,900 years. None of these are constant, pure frequencies. In fact, the relative postions of the earth, sun, and moon are never exactly the same.

On the average, the lunar tides are equal to $5.7 \times 10^{-8}g$ or about 50 μgals (a gal is 1 cm/s^2) and the solar tides, 0.46 times this or 23 μgals. The tidal force at any particular point at any particular instant is determined by two external gravitational fields, one due to the sun and the other to the moon, whose absolute and relative amplitudes will differ with time.

By far the largest variations in tidal force are caused by the rotation of the earth, occurring every 24 hours and generating the semidiurnal and diurnal tidal components. Due to the 23.5° tilt of the earth's axis with respect to the normal to the plane in which it orbits, these usually are not equal; the inequality varies as the moon revolves around the earth.

The longer tides modulate the amplitudes of the daily tides. In general, the moon moves around the earth in about 28 days in a somewhat elliptical orbit. As the moon's motion is very complicated, actually the rate of revolution varies. It does not have a pure 14 day period, varying from 11 to 16 days. The maximum tides occur when the sun and the moon are pulling together in conjunction and in opposition, at new moon or at full moon; the minimum, when the moon is in quadrature. The fortnightly variation in the tidally induced gravity is about 100 percent, ranging from 95 to 200 μgals at 45° latitude of Yellowstone.

The semiannual component, resulting from the 23.5° tilt of the earth's axis with respect to its orbital plane, is caused directly by changes in the angle of the radius vector during the earth's yearly trip around the sun. The tides are the highest at the winter and summer solstices, 21 December and 21 June, respectively. The variation ranges from 10 to 15 percent.

A 5° inclination of the orbital plane of the earth with respect to the orbital plane of the moon causes significantly large and periodically varying changes in tidal potential. These periods are 4.4 years, 8.8 years, and 18.6 years. The amplitude of the 18.6 year component of tidal gravity is the largest of the three, about 10 percent of the maximum daily variation. The 4.4 and 8.8 year components interact with the 18.6 year component, producing abnormally high and abnormally low variations that occur every 50 years or so. The magnitudes and variations of the amplitudes of all the components are strongly latitude dependent.

Tidal forces are observed to have strong regulatory effects on geyser activity even though the strains produced by these forces acting on competent rocks, reacting elastically, would be only of the order of 10^{-7} or 10^{-8}, certainly insufficient strain to cause any appreciable effect. The rock mass must therefore react inelastically. A possible model is that shown in Fig. 8-12. When the tidal force is lowest, the effect of gravity is the greatest, tending to pull the rocks downward and compressing the pores and rubble-filled fractures. When the tidal

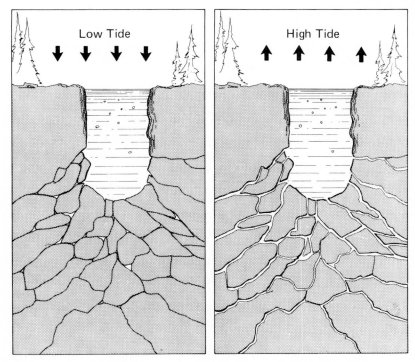

Figure 8-12. Effect of variation in g by tidal forces in modifying crack openings.

force is high near the times of new and full moon, the rocks and gravels tend to float upward on the sheets of saturated material on which they rest, opening up the channels and permitting more hot water to flow into the geyser reservoir.

Such a notion is supported by the observation that the semidiurnal and diurnal tides, even though by far the largest, have no evident effect on the geysers. However, the fortnightly tide which acts over a longer time, produces substantial inelastic deformations and does affect the geysers. Riverside's reaction to it is striking, its approximate 6 hr interval fluctuating plus or minus 15 min in sympathy with the tide. Figure 8-13, a plot of intervals between eruptions vis-á-vis a plot of variation in gravity due to tidal forces, illustrates the close correlation that existed during the summer of 1967. The same correlation exists in other years, sometimes less and sometimes more prominently. An increase in gravity causes the geyser to erupt more frequently. During May and early June and again from July to October the correlation between the eruption interval and maxima in earth tidal forces is very high. The fortnightly component is especially evident when this component is relatively large but fades out in September and October when it is obscured by the semiannual component. During late June and early July the buildup of stresses anticipatory to the occurrence of two sizable, 4.2 and 4.0, earthquakes in late July adversely affected the correlation. The two to three day displacement between the troughs of the interval curve and the peaks of the tidal

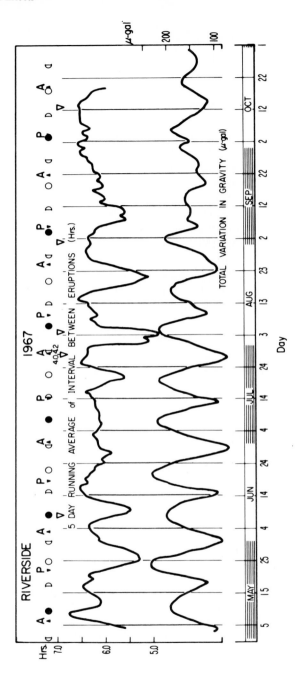

Figure 8-13. Activity of Riverside Geyser vis-á-vis variations in earth tidal forces.

force curve implies a short phase lag between the time when the tidal force acts and the geyser responds.

Old Faithful also responds to the fortnightly tide, similarly exhibiting a two to three day time delay and being further perturbed by the buildup and release of tectonic stresses associated with earthquake activity. A quantitative assessment of the effect of gravity variations alone is shown in Fig. 8-14, a plot of interval between eruptions versus the maximum tidal force. Each plotted point is the average of a day's readings, all taken during the summer of 1968 when the fortnightly component was especially strong.

Grand and Steamboat Geysers respond dramatically to the 18.6 year tidal component. The intervals between eruptions at Grand vary seasonally from 8 hr to over 40 hr, with two periods of dormancy over the four decades of observation. The high correlation between length of interval and tidal force is obvious from a visual comparison of the two curves on Fig. 8-15. During times of high tidal force, around 1930, 1955, and 1970, Grand erupted two to three times daily, only to become almost dormant during the years 1943 and 1960. Late in the 1970s, its interval began to increase again as the tidal force lessened.

Steamboat's behavior for the much shorter period 1963 through 1978, has been just the opposite, its tremendous eruptions tending to be triggered by the low tidal stress.

The difference in the behavior of the two geysers, Grand becoming less active

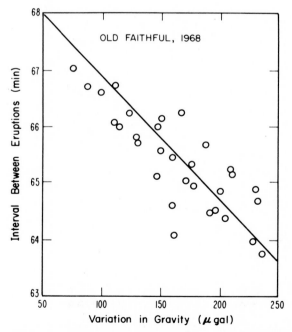

Figure 8-14. Average interval between eruption of Old Faithful plotted against variations in tidal force, mainly attributable to the fortnightly component.

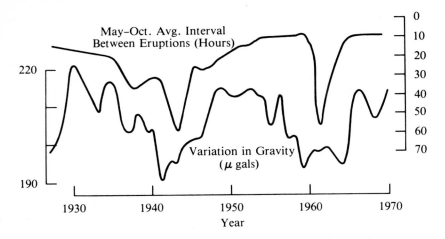

Figure 8-15. Eruption frequency of Grand Geyser vis-á-vis variation in average earth tidal force due primarily to 18.6 year component.

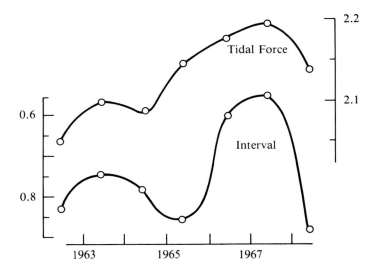

Figure 8-16. Interval between eruptions of Old Faithful of California Geyser vis-á-vis variation in average earth tidal force. The 4.4 year component is the most prominent one.

during times of low tidal forces and Steamboat more active, may be due to the fact that they lie in distinctly different types of geothermal areas. During 1971, a time of relatively high tidal force, Steamboat was dormant. But the pools and mud pots of Norris Basin became exceedingly active, suggesting a change in the pattern of circulation. Although the total energy output increased, energy was sapped somehow from Steamboat to feed the pools. On the other hand, Grand

decreases its activity at low tide when nearby Rift Geyser, normally inactive at times of high tide, becomes active and saps Grand's strength already enervated by the general decrease in energy flow.

Old Faithful of California's reaction to the long period tides over about a seven year period is shown in Fig. 8-16. Although less pronounced, the response is substantial, increasing gravity causing an increase in frequency of eruption of the geyser.

CHAPTER 9

Man's Influence on Geyser Activity

9.1 Some Man-Made Geysers

A few of the many wells that have been drilled in thermal areas for various purposes have turned into geysers; some are only short-lived but others have stayed active for decades. These geysering wells are ideal for studying the basics of geyser activity, especially when detailed records of the geologic and thermal features of the well and surroundings were made at the time it was drilled. Such studies are often more informative than those on natural geysers whose detailed plumbing systems are unknown, and except in the uppermost reaches of the vent, inaccessible for making temperature and other measurements. Besides this, wells can in most cases be freely regulated and tampered with.

The Japanese have been particularly active and successful in developing many new geysers by drilling. Some of them, with their eruption characteristics shortly after drilling, are listed in Table 9-1. All told, six artificial geysers, Miyazawa and Benten Nos. 1-5, were developed at Onikobe in the mid 1930s by drilling close to a small natural geyser that had become quiescent about 20 years earlier. Extensive observations were made of their activity from 1940 to 1950. The behavior of all of the geysers changed significantly during this period: their intervals between eruptions, their eruption patterns, and the duration of play. Toward the end of the study, all of them had become less active.

Measurements made down within many of the wells indicated that temperature regimes were about the same from well to well and varied about the same way with depth. Before an eruption with a well full of water, the temperature in Benten No. 5 in the depth range 10 to 16 m was 116°C ± 1°C. A depth of 15 m appeared to be critical. Deeper than this, at 17 m, cooling did not begin for about 2 min after an eruption started whereas at shallower levels, cooling began immediately. The rapid drop in temperature shortly after an eruption starts is simply due to water beginning to boil as pressure is reduced as a result of discharge of

Table 9-1. Characteristics of Japanese Artificial Geysers (Adapted from Noguchi, 1956)

Geyser	Interval between Eruptions	Length of Play	Depth of Hole (m)	Height of Eruption (m)
Miyazawa	22 hr	1 hr	—	—
Benten No. 1	11.8 min	0.42 min	18.5	2
Benten No. 2	1.8 min	0.25 min	16.0	15
Benten No. 4	16 hr	1.44 min	30.7	25
Benten No. 5	5 hr	10 min	17.6	20
Tosenro (large)	7 hr	2 min	—	—
Tosenro (small)	7 hr	5 hr	—	—
Kanetyu	4.8 hr	13.8 s	—	—
Tutiyu	35.5 min	1.4 min	—	—
Tatumakizigoku	10–18 min	2–8 min	—	—
Yunodanc	4–6 hr	~30 min	—	—
Sikake	8.7 min	2.8 min	—	—

water from the top. After cooling, the water at the 17 m level began to heat up slowly except for a few downward excursions, becoming hot enough to erupt in about 5 hr. The general character of the temperature variations were substantially the same at all levels, although at the very shallow levels temperatures ran a few degrees lower.

At any particular time, the temperature from the 15 to 7 m depth did not change with depth, a condition adequately accounted for by assuming the existence of a cavity created subsequent to drilling of the well in which convection could take place freely and thus eliminating temperature gradients. A subsurface cavity of this type could easily develop as a result of sloughing of the uncased walls of the drill hole prompted by the vigorous action of rushing water and steam. Presumably steam collects near the roof of the cavern and eventually develops sufficient pressure to force enough water out of the well to precipitate an eruption.

Many observations were made in the late 1940s and early 1950s on the activity of an abandoned well, No. 32, at Steamboat Springs. Although the well was reportedly drilled to a depth of about 20 m, only the upper 13 m was accessible for the study. Presumably the remainder been filled in with rocks, sand, and gravel. It was cased with a 12 cm diameter iron pipe to a depth of 10 m. At the time of the study, major eruptions occurred at intervals of five to seven days, throwing water to maximum heights of 18 to 23 m for 30 to 40 min. Water level immediately before an eruption never reached ground level, staying 12 to 80 cm below. A violent steam phase lasting for about 5 min followed an eruption which in turn was followed by a passive steam phase lasting about 20 min.

Measurements between eruptions made down in the well indicated that in general temperatures at all levels were always at least 3°C below the boiling point curve for pure water. Considerable convection seemed to be taking place, leading

to the formation of many bubbles as blobs of hot water rose to regions of less hydrostatic pressure where the water could transform into steam, eventually precipitating an eruption. The total measured discharge of water during an eruption was 6500 liters. Since the volume of the geyser tube was only 200 to 250 liters, an additional reservoir probably existed somewhat as indicated in Fig. 9-1, a cross sectional view of the well with an hypothesized reservoir. This may be somewhat similar to the supposed additional reservoir at Benten No. 5. Addition of fluorescein as a trace dye suggested that the reservoir could be as large as 400,000 liters, only partially emptying during an eruption.

Although no written records have been found regarding its drilling, Old Faithful of California presumably is a drilled well that geysers. According to old settlers of the area, the driller was nearly blown to bits when the well began erupting while he was drilling. The hole is 70 m deep and cased part way down with a 30 cm diameter pipe. It erupts about every 40 min to a height of 20 m, playing for 2 to 3 min.

It sits in a sedimentary basin where considerable drilling has been done to enhance the local water supply and to provide hot water for health, recreational, and agricultural purposes. Although it has been erupting more or less regularly for the past 80 years, it fills up with debris every few years, closing off the casing at about the 13 m level. Reaming restores it to normal operation. Once vandals stopped the geyser from operating by jamming its opening with empty glass bottles.

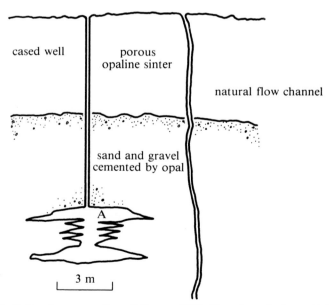

Figure 9-1. Inferred cross section of Geyser Well, Steamboat Springs. (Adapted from White, 1967.)

In 1935, a rancher in the Crump Lake hot springs area of southeastern Oregon drilled a 13 m deep well for water for his cattle. Twenty years later he heard about others developing geysering drilled wells and decided to follow suit. He was eminently successful when he deepened his well to 35 m. Since then the well has continued to erupt to a height of about 30 to 40 m on an 8 to 10 hr schedule, playing for about 1 min. Each such eruption is followed within a few minutes by a somewhat lesser eruption, and occasionally by a third. The one or two eruptions following shortly after the main eruption most likely result from water entering an already preheated tube, which causes it to heat up quickly to an unstable situation. This is a much exaggerated case of bimodal distribution in intervals observed in numerous natural geysers.

The well is steel-cased to a depth of about 12 m, the upper 3 m being 20 cm casing and the lower 9 m, 15 cm casing. It is topped with a rectangular wooden box 40 cm deep and 1 m on a side. After an eruption, the tube fills relatively quickly, the rise being about 15 cm/min. This rate of flow is maintained until the next eruption. The excess water spills over the edge of the box.

In an effort to produce steam for generating electric power, a 500 m deep well was put down close by. It shot hot water and steam continuously 70 m into the air for ten months until vandals plugged it by bulldozing the dry walls of an old stone corral into it.

At Lakeview, Oregon, about 60 km to the west of Crump, are two drilled wells. One is used for space heating a motel. The other erupts almost perpetually, every 15 s for 8 s.

9.2 Changing the Activity of Geysers

Achieving a preselected and desired change in a geyser's performance is easy when the characteristics of its heat and water supplies and reservoir system are known. Such knowledge was used nearly a century ago to make Pohutu Geyser erupt more often. Pohutu is interconnected with Prince of Wales Feathers, Te Horu, and Waikorohihi. Before steps were taken to change the situation, Waikorohihi erupted almost continuously and discharged water into Te Horu. Pohutu's eruptions were spaced at a several month interval. The first modification was to build a wall around Waikorohihi, preventing it from discharging into Te Horu. As a consequence, Te Horu raised its water level about 1 m and started boiling vigorously. This caused Prince of Wales Feathers to become active, at which time boiling in Te Horu stopped and the water level fell. The next step was to quench Prince of Wales Feathers by directing down it a stream of cooled water from Waikorohihi's overflow. Te Horu became active again and shortly thereafter Pohutu began erupting about every 12 hr, continuing to do so for a few years until the "improvements" fell into disrepair. In recent years, Pohutu and its interconnected geysers have become active again in a modified pattern.

A favorite and embroidered Yellowstone story concerns a Chinese laundryman who arrived in the 1880s to take up business in the Upper Geyser Basin.

Loading a nearby hot pool with dirty clothes and soap, he caused an eruption that tossed his tent, belongings, and laundry high and wide. Thus, Chinaman Spring.

Soap is frequently used to trigger the action of a geyser, especially by entrepreneurs anxious to satisfy an impatient tourist. Lady Knox, a small privately owned geyser in New Zealand, is advertised to erupt at 10:30 each morning. About 9:30, the tourists assemble, the owner plunges a few kilograms of soft soap down its vent and covers the opening with some cloths which he removes when in a few minutes the vent begins to froth. Gradually a foam-filled column rises to at most 30 m for several minutes.

It was a common practice about the turn of the century for the Maoris at Whakarewarewa in a solemn ceremony to soap Wairoa, its largest geyser, for the benefit of a distinguished visitor. Natural eruptions rarely occurred. The ceremony began by careful removal of the usually padlocked cover from the mouth of the geyser. After the crowd had been warned to stand back out of danger, a bag of bar soap was thrown into the small opening. Almost immediately the water foamed, lathering up to the edge of the sinter opening, but it was 20 minutes before the geyser actually started playing. The eruption began with rumblings, gurglings, and choked noises coming from depth, followed by a great roar that shot a column of white steam and water 60 m into the air. The usual eruption lasted a minute or so; the longest observed one played for more than 2 hr. Soaping eventually became unsuccessful and was stopped.

The main action of soap is to produce a very large number of bubbles that by displacing the water reduce the hydrostatic head, allowing the water at lower levels, already superheated with respect to the ambient boiling point at the surface, to flash into steam and precipitate an eruption. The bubbles form principally because soap, even in small concentrations, drastically lowers the surface tension of water, even as much as 60 percent. At first, there are only a few large bubbles formed. Their growth is checked by evaporation of water into them which leaves a film of soap surrounding them. The insoluble soap of the film serves as numerous nuclei for absorption of dissolved gases and for the initiation of large numbers of new bubbles.

An unobstructed discharge vent greatly enhances a geyser's ability to perform. An owner of a geyser, especially if he is displaying it to tourists, keeps the geyser reservoir free of debris. A number of years ago, a continuously boiling hot spring at Hveragerdi was transformed into a vigorous geyser by removing loose stones and other obstructions from the accessible parts of the conduits, permitting hot water and steam to freely rush out. In addition a nearby merchant sold a ''special'' soap to further facilitate the action. A number of presumably dead geysers have rejuvenated themselves when some subterranean instability generates a flow of steam and hot water sufficiently powerful to scrub the vent free of debris. Contrariwise, a geyser's eruptions can often be stopped by filling its vent with debris. This has been done where the action is annoying or dangerous to the environment. Sometimes the filling occurs naturally. The present inactivity of the Great Geysir is believed by many to be due to obstacles in the filling channels.

One common way to prod a geyser into activity is to lower the water level in

its reservoir which is easily accomplished by digging a trench or boring a horizontal hole in the cone. Lowering the hydrostatic head allows the superheated water at lower depths to boil. There are many such examples. By lowering the water level only 60 cm in the boiling spring Pura, it was transformed into a geyser throwing water 9 to 12 m into the air. At one time Uxahver was playing about 3 m high about every fifth minute. When the water level was lowered 25 cm by making a cut in the lip of the basin, it played more often but less vigorously, erupting every 2 to 3 min to a maximum height of only about 2 m.

At one time eruptions of Hokozigoku Geyser in Japan were regulated by raising and lowering the water level in its basin. Raising the water level lengthened the interval between eruptions with the maximum temperature attained by the water being lowered. Lowering the level shortened the interval and raised the maximum water temperature. The amount of change depended on the change in water level. Lowering the water level by about 37 cm decreased the interval from 52 min to about 29 min.

Solitary Geyser only became active when it first began to be used as a source of hot water for baths and a swimming pool at a local tourist hotel. The approximate 1 m lowering in water level apparently lowered the pressure sufficiently to induce boiling below the surface.

One enterprising Icelander transformed his ordinary hot spring, Bogi I, which lay on the side of a hill, into a geyser by drilling horizontally in from the side and installing an iron pipe equipped with a valve (Fig. 9-2). With the valve open, the water stood at level B and the spring functioned as a geyser erupting on a 10 min schedule playing for 2 to 3 min to an 8 m height. With the valve closed, the water level rose to level A where it continuously boiled vigorously but did not erupt. When boiling vigorously, the water at A was superheated to 103°C but when functioning as a geyser, the temperature of the water at B was only 100°C.

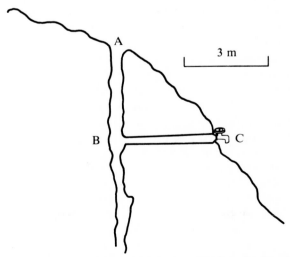

Figure 9-2. Cross section of Bogi I showing man-installed lateral valve. (Adapted from Barth, 1950.)

The Great Geysir is perhaps the most famous geyser to be rejuvenated by lowering the water level. The geyser became dormant in 1915 and remained so until 1935. A narrow gate was cut in the rim of its basin, which, when open, lowered the water level about 60 cm. As a usual thing, the gate was kept closed until an eruption was wanted, generally on Sunday for tourists. A large eruption could then be precipitated almost immediately by opening the gate and then throwing in several kilograms of soft soap. When the gate was left open after such a large eruption so that the tube could refill only to the level of the gate, small eruptions occurred at intervals of several hours without the addition of soap.

Eruptions are sometimes precipitated by blowing air into the filled geyser vent and causing the water to overflow. This lowers the pressure on the water below, allows more hot water to flow in, and, if the water is hot enough, initiates an eruption. This was regularly successfully done in wells at Steamboat Springs using bottled compressed air fed through a tube. At the Tosenio geysering well in Japan, air was simply blown in from a rubber tube introduced into the orifice.

A considerably more unusual and spectacular way to precipitate an eruption is to toss in a chunk of dry ice. The effect of its rapid production of carbon dioxide vapor in the hot water greatly overshadows its cooling effect.

Release of gas can also be stimulated by throwing handfuls of sand or gravel and even larger pieces of foreign matter such as clods of earth into the water. Striking the waters with ropes and tree branches or by stirring them up with a pole is equally as effective. This is because generally the waters coming up to shallow levels from great depth where the solubility of gases is high under the considerable hydrostatic pressure, release the gas only slowly under the reduced pressure unless disturbed.

Another very effective way used to initiate an eruption is to transport some of the superheated water existing at depth suddenly to the top of the geyser vent where it flashes into steam. Lowering the hydrostatic pressure allows the remainder of the lower waters to flash. A common method is to lower a bucket weighted with stones and to which a rope is attached down into the vent where it fills with superheated water. The bucket is then quickly pulled up, and bucket, stones, rope, and water all erupt.

Controlled experiments aimed at modifying the performance characteristics of geysers proved very useful for determining the conditions leading to geyser eruptions. Russian scientists have systematically studied two of the geysers, Quathegey and Prince Buratino, located in Geyser Valley, Kamchatka. Both are cone geysers with small pools surrounding them. They are both characterized by a distinct preliminary discharge stage and by an interval between eruption not varying more than 4 percent. Most of the ejected water runs off into the Geysernaya River on whose banks they sit. Quathegey's 10 cm diameter exit vent is vertical to a depth of 1.5 m below which its orientation changes. Prince Buratino has a 2 to 4 cm wide, about 20 cm long fissure-like exit vent inclined at 50° to 60° with the vertical. Important characteristics of the geysers are tabulated in Table 9-2.

The experiments consisted essentially of pouring various known quantities of

Table 9-2. Main Phases of Activity of Quathegey and Prince Buratino Geysers (Adapted from Steinberg et al., 1978)

Phase	Quathegey	Prince Buratino
Full period of geyser from start of one eruption until start of next	12.8 min	9.3 min
Duration of filling of geyser from end of eruption until beginning of preliminary discharge	10 min	6.1 min
Duration of preliminary discharge stage	0.8 min	1.5 min
Duration of eruption	1.8 min	1.8 min
Volume of water thrown out during eruption	120 liters	120−150 liters
Full free volume of system after eruption	180 liters	180 liters
Maximum height of water	1.5 meters	2.5−3 meters
Natural rate of inflow of hot water immediately after eruption	0.53 liters	0.61 liters
Average heat energy added by inflowing hot water and steam	164−174 kcal/kg	180−185 kcal/kg
Upper level of steam pressure	8.4 atm	10.8 atm

different temperature water into the geysers and observing the effect the added water had on the intervals between eruptions. The amount of water added immediately after an eruption ranged from 180 liters, the total amount each of the geysers would hold, down to 10 liters. Water temperatures ranged from 80°C to 10°C. From these data the following were then calculated:

- average heat content or enthalpy and the rate of flow of the fluids naturally entering the geyser system;
- variations in the process of filling the system; and
- the amount of unfilled reservoir volume existing after an eruption.

The data showed that the addition of small amounts of water, 10 to 40 liters, shortened the interval between eruption, with large amounts of water, 80 to 180 liters, increasing the interval, more or less linearly with amount of water added. These two distinctly different behavior patterns are primarily the result of interplay between hydrostatic pressure regimes and temperature regimes as the added and inflowing waters flow into respective parts of the geyser reservoir system.

Practical Uses of Geothermal Fluids

10.1 Introduction

Geothermal areas, geyser areas, are scenically gorgeous. But they are also an important energy resource, largely waiting to be tapped. There has been an extensive development in locating and efficiently exploiting the tremendous heat of geothermally active regions to generate electric power, heat homes, and operate chemical plants. Exploitation of geothermal heat has occurred ever since man first warmed himself by a hot pool. The hot baths of Roman times and the traditional use of thermal springs in Japan through the centuries have been largely for recreation and health. But beginning about the turn of this century, the use of geothermal resources has been expanding and now represents a much needed alternative to fossil fuels.

For commercial exploitation, a geothermal reservoir must meet several exacting standards. It must have an adequate heat source with a geothermal fluid at a temperature suitable for the application at hand: high, 150°C, for power generation, or considerably lower, as low as 40°C for space heating, and some industrial, agricultural, and recreational purposes (Table 10-1). It must be obtainable at a relatively shallow, 3000 m or less, depth. The reservoir must be large and sufficiently permeable to allow water or steam to flow out continuously at a high rate, or be stored. Yet for many uses, it must have capping rocks of low permeability which inhibit the flow of water and heat to the surface. Sufficient water is necessary to recharge or maintain production over many years. And it must be relatively free from excessive and corrosive chemicals.

The transportability of geothermal fluids, which is relatively low, is highly dependent on the temperature of the fluid. Several installations in Iceland move 100°C water as far as 10 to 20 km by pipeline; 150°C to 180°C water can be efficiently transported 50 to 75 km for residential heating purposes.

Table 10-1. Temperatures of Geothermal Fluids Required for Various Purposes (Adapted from Lindal, 1973)

Degrees (0°C)	Use	
180	Evaporation of highly concentrated solutions; Refrigeration by ammonia absorption; Digestion in paper pulp, kraft	
170	Heavy water via hydrogen sulphide process; Drying of diatomaceous earth	
160	Drying of fish meal; Drying of timber	
150	Alumina via Bayer's process	
140	Drying farm products at high rates; Canning of food	
130	Evaporation in sugar refining; Extraction of salts by evaporation and crystallization	
120	Fresh water by distillation; Most multiple effect evaporations, concentration of saline solution	
110	Drying and curing of light aggregate cement slabs	
100	Drying of organic materials, seaweeds, grass, vegetables, etc.	
90	Drying of stock fish; Intense de-icing operations	
80	Space heating; Greenhouses by space heating	
70	Refrigeration (lower temperature limit)	
60	Animal husbandry; Greenhouses by combined space and hotbed heating	
50	Mushroom growing; Balneological baths	
40	Soil warming	
30	Swimming pools, biodegradation, fermentations; Warm water for year around mining in cold climates; De-icing	
20	Hatching of fish; Fish farming	

Markings at left: *Saturated Steam* (spanning approximately 100–180), *Water* (spanning approximately 20–120). Marking at right: *Conventional power production* (spanning approximately 130–180).

There are five general types of natural geothermal resources that can be put to practical use:

- hot, 60°C to 100°C water that flows from natural or man-made holes at a temperature less than its boiling point (a hot water system);
- wet, saturated steam which is formed when superheated water gushing from drilled holes partly flashes into steam (a water-dominated system);
- dry steam which streams from holes drilled to great depth (a steam-dominated system);
- geopressurized reservoirs in which hot fresh water at high pressure occurs in natural deep underground reservoirs; and
- hot rocks, deeply buried, from which dry steam can be generated by dousing with water.

Although electric power generation will probably remain the most important application for geothermal energy, it is being increasingly used for space heating, for chemical extraction, for raising forced vegetables and flowers in greenhouses, for heat-intensive industries such as paper making in New Zealand and drying of diatomaceous earth in Iceland. Borox has been produced geothermally at Larderello since the eighteenth century. Perhaps the most pressing need in the coming years will be adequate fresh water. Geothermal desalinization in certain areas such as the Imperial Valley in California holds great promise. The locations, approximate sizes, fluid temperatures, and uses of several exploited geothermal areas are listed in Table 10-2. Table 10-3 details the worldwide exploitation of geothermal resources as of 1975.

Japan is perhaps the leading nation geothermically speaking. It has between 13,000 and 14,000 geothermal wells. Because it lacks indigenous fossil fuels, Japan has always utilized geothermal fluids most particularly for agriculture and tourism. More than 150 million Japanese enjoy therapeutic and recreational facilities.

Iceland has 17 known high temperature geothermal fields and low temperature fields are scattered throughout the country. The energy is used primarily for space heating.

In New Zealand, it is used for industrial processing, space heating and cooling, and generating electric power. Italy has been generating electric power since the early 1900s and the USA and states in Latin America are following suit.

In eastern Europe, Bulgaria, Czechoslovakia, East Germany, Hungary, Poland, and Roumania all utilize hot waters obtained from high flow areas for space heating, agricultural, and industrial purposes. In the USSR, lower temperature resources, 47°C to 86°C, located near the Caspian and Black Seas are used for space heating, agricultural, industrial, and recreational purposes.

Promising exploratory programs are underway in Israel, Turkey, Ethiopia, and Kenya.

Table 10-2. Characteristics of Several Exploited Geothermal Areas (Adapted from White, 1965)

Area	Approximate Size (km²)	Temperature (°C)	Use
Iceland			
Reykjavik	~5	146	space heating
Namafjall	2.5	—	electric power
Italy			
Larderello	~50	240	electric power
Japan			
Otaki	—	185	electric power
Onikobe	80	185	electric power
Beppu	10	150	electric power
Matsukawa	—	189	electric power
Mexico			
Cerro Prieto		380	electric power
Pathé	2	155	electric power
New Zealand			
Wairakei	7	266	electric power
Rotorua	—	>160	space heating
Kawerau	—	285	industrial processing
United States			
Brady's Springs	~2	187	food processing
The Geysers	~1	208	electric power
Imperial Valley	—	—	electric power, desalinization
USSR			
Pauzhetka (Kamchatka)	~1	195	electric power

Table 10-3. Worldwide Exploitation of Geothermal Energy (Numbers are Thermal Mw) (Adapted from Peterson and El-Ramy, 1976)

Country	Agricultural	Balneology/ Tourism	Electric	Space Heating	Industrial	Total Percent
Japan	340	629	43	28	57	32
USSR	234	116	6	71	11	13
USA	6	<1	396	8	—	12
Italy	1	—	405	—	—	12
Hungary	125	232	—	10	21	11
Iceland	40	—	3	254	18	9
New Zealand	—	—	202	32	43	8
Mexico	—	—	75	—	—	2
Total	745	978	1130	404	150	3407
Percent	22	29	33	12	4	—

10.2 Characteristics of Exploited Geothermal Areas

Most of the geothermal areas where energy is commercially exploitable differ geologically from the principal geyser basins in that natural surface flow is prevented by an impermeable cap rock which must be penetrated in order to extract hot fluids. Having little or no surface expression, they are hard to locate. Drilling is still the surest and most effective exploration tool but it is expensive. At first, infrared photography was thought promising but studies have shown that it does not differentiate sufficiently between surface and subsurface thermal features. Many thousands of heat measurements have been made and while above normal heat flow indicates high temperature gradients, suggestive of potential reservoirs, they are often misleading as regards their extent, position, and thermal characteristics. Wet rocks, being much better electrical conductors than dry ones, exhibit easily identifiable electrical resistivity anomalies that can be located by well-established geophysical prospecting techniques but often they do not uniquely define the source. Gravity and magnetic surveys are successful only to a limited extent, mostly in complementing other types of surveys.

The new technique of ground-noise mapping appears to be the most promising. A geothermal area has a characteristic set of ground noises and micro-earthquake activity. The technique consists of using geophones capable of sensing ground vibration to survey a suspected geothermal area and establish the spatial and temporal distribution, frequency content (of the order of 1 Hz) and amplitude of the motion of the surface of the ground. Using the ground vibrations of recognized geothermal basins as baselines, other fields can be delimited. The locations and geologic characteristics of many of the geothermal areas now being exploited are listed in Table 10-4.

The essential requirements of an exploitable area are relatively high temperatures and permeable and faulted structures that can yield heat-transporting fluids

Table 10-4. Reservoir Geology of Some Geothermal Areas (Adapted from Banwell, 1973)

Field	Fluid	Source Rock
Larderello, Italy	steam	fractured limestones; dolomite
New Zealand	hot water	acid volcanoes
Japan	hot water	acid volcanoes
The Geysers, Calif.	steam	fractured graywacke
Cerro Prieto, Mexico	hot water	river delta sediments
Imperial Valley, Calif.	hot water	river delta sediments
Pathé, Mexico	hot water	fractured middle Tertiary volcanics
Iceland	hot water	fractured cavernous basaltic lavas

to a well. Usually the major areas occur in regions of normal faulting caused by tectonic tension and accompanied by upward flow of magma (Fig. 2-4). The geothermal areas of Kamchatka, USSR, Matsukawa, Japan, and Valle Grande, USA, are all located on the rims of calderas and are controlled by the intersection of fissure zones. Otake, Japan, Matsas, Taiwan, and Boiullante, Guadeloupe, are located on volcanic domes. The Wairakei, Broadlands, Kawerau, and Rotorua fields of New Zealand all lie along the Taupo Volcanic Zone (Fig. 1-10); the Imperial Valley and Cerro Prieto fields are related to the crest of the East Pacific rise; Reykjanes and Namarfjall are located along the crest of the mid Atlantic Ridge; and The Geysers, Sulphur Bank, Steamboat Springs, and Brady's Springs, USA, El Tatio, Chile, and Ogaria, Kenya, are all located in rifts or spreading basins. Larderello, Italy, Kizildere, Turkey, and Surprise Valley, Calistoga, and Beowawe, USA, are in areas that are spreading but exhibit no very recent volcanism.

At some geothermal areas faulting does not play a role. Steep fissures are scarce and heating is done by conduction. Temperatures, less than 100°C, in this type of field are lower than in fault block areas. Examples are the Hungarian Basin, Hungary; the Neovolcanic region, Slovakia; and the Intermountain Basin of Georgia, USSR. Some field simply exhibit average heat flow and geothermal gradients. Some of these are located in Czechoslovakia, Poland, and the Crimea, West Siberia, and the Turana Lowland areas of USSR and are used mainly for balneology, greenhouses, and dairy farms.

The respective geologic nature of the important fields listed in Table 10-4 are seen to differ. Whether a field will produce wet or dry steam or hot water is partly controlled by how much the thermal fluid is diluted by cold, near-surface ground water which in turn is controlled by the structure and permeability of the rocks underlying the thermal area. When the fissures conducting the hot fluid are situated in a structural depression filled with porous sediments and volcanic debris as in the Imperial Valley, the ascending hot fluid will mix with cooler ground water and the resultant mixture will saturate the porous surface rocks. In this case, the geothermal area is water-dominated and produces saturated steam and hot water. When the fissures intersect only impermeable rocks as at The Geysers or if they are protected by an impermeable layer from downward percolation of near surface ground water, as in Italy, surface water will only slightly dilute the thermal fluid. Then the field is likely to yield dry superheated steam depending upon its initial heat content.

The two resources, The Geysers and Larderello, which produce most of the electric power, about 500 Mw each, are steam-dominated. Other large power producers, Wairakei and Broadlands, and Cerro Prieto are water-dominated. Any exclusive classification is not clear-cut; one type blends into another. In fact, one of Wairakei's producing fields changed from a water-dominated system to a steam-dominated one as the flow from the wells exhausted the water flow.

The Geysers yields dry steam at 7 kg/cm^2 pressure and 205°C temperature from depths ranging from 600 to 3000 m. The thermal characteristics of the Larderello reservoir are similar. The Geysers lies in the Mayarinas Mountains of

northern California. The rocks in the area are a sequence of graywacke (quartzone, sandstone, and argillite), shale, basalt, and serpentine which is uplifted and completely faulted into a series of northwest-trending horsts and grabens, ranging between 2 and 4 km in width, and retaining their identity for lengths up to 15 km. The Geysers occurs along the canyon of Big Sulphur Creek. Here the serpentine and basalt have been downdropped into the underlying graywacke. The thermal areas lie in the fissures closest to the serpentine body, the stratigraphic section that has undergone the greatest amount of subsidence (Fig. 10-1). The original natural thermal springs occur mainly on the southwest side of the serpentine, probably reflecting the pattern of ground water flow since this side is considerably lower in elevation.

The rock sequence at Larderello, on the other hand, is entirely sedimentary with no igneous rocks evident. Steam production comes from highly pervious carbonate and anhydrite of upper Triassic to upper Jurassic age. The basement, a chaotic thrust complex of clay, limestone, and other materials is faulted, allowing hot fluids to flow upward into the anhydrite. Impermeable clays form a cap rock that seals in the fluids and heat of the geothermal system. Deep faulting forms an intersecting network that localizes the hydrothermal activity. Largely steam and gas rather than hot water, flow from the natural outlets of the area.

The water-dominated Wairakei and Broadlands geothermal resources are collocated with some of the geyser fields in the still volcanically active Taupo tectonic depression. A vertical geologic section at Wairakei begins at the surface with 100 m of soft rock and pumice, followed in downgoing order by 35 m of impermeable dense fine sandstone and mudstone, 500 m of porous cemented and fissured pumice and rhyolite, and finally, massive ignimbrite of indeterminate thickness (Fig. 10-2). The porous and fractured rhyolite provides the reservoir for the storage of water and steam. The main channels for the upward flow of hydrothermal water are the subsurface extensions of major active faults such as the Wairoa and Wairakei.

The geology of Broadlands is similar. It consists of a stratified volcanic sequence resting on a graywacke basement. Major thermal anomalies with tem-

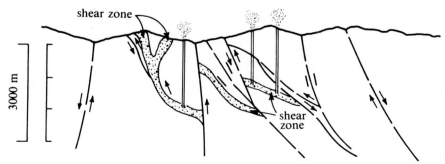

Figure 10-1. Cross section of The Geysers area showing fault shear zone, and drilling pattern. (Adapted from Anderson, 1970.)

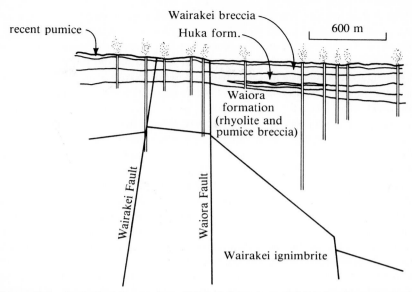

Figure 10.2. Geologic cross section of Wairakei geothermal field showing some typical producing wells. (After Grindley, 1970.)

peratures exceeding 300°C lie along buried fault-block horsts. Four aquifers are interspersed among two extensive impermeable rhyolite flows and three thick impermeable ignimbritic ash flows. Drill holes are sited along faults transecting a rhyolite dome that caps one of the aquifers.

Typically, Iceland is built up of inhomogeneous layers of Tertiary basalt reaching thicknesses of 3 km or more. In some areas, these layers are overlain by Quaternary basalts containing recent faulting and areas of hot water with temperatures less than 150°C. The Quaternary areas lie on a major northeast-trending rift zone that bifurcates to the southwest. Springs with subsurface temperatures ranging from 150°C to 200°C, natural steam vents, and areas of thermal metamorphism lie along the rift proper. Specific thermal areas are controlled by intersections between permeable volcanic layers, dikes, and faults. Water circulation systems are believed to almost contact very hot magmatic intrusions. Drilling indicates that large amounts of high temperature water can be produced from shallow depth, less than 450 m. Deposition of silica and calcite have formed an almost impermeable near-surface layer several hundred meters thick which restricts circulation.

The Mexicali-Imperial Rift Valley is one of the largest geothermal provinces in the world. The valley is a structural depression extending from the head of the Gulf of California northwestward through the Colorado River delta and Salton Sea to the Coachella Valley. The Imperial Valley portion lies in the USA and is 120 km wide and 160 km long, with the granitic Peninsular Range bounding it on the west and the Chocolate Mountains, a complex of Precambrian igneous and metamorphic rock capped by Tertiary volcanic rock, bounding it on the east.

The structure is unique, having predominantly transform faults and lateral movement. The rift floor consists of a fault-cut prism of water-saturated sediments which reach a maximum thickness of 6000 m. The sediments are mainly continental deposits of fanglomerate, conglomerate, and lacustrine sandstone and claystone deposited by the Colorado River and other streams during the last 4 million years. They contain a reservoir of up to 1 billion hectare-meters of very hot ground water, insulated from the surface by impermeable silts and clays. Drill cores show that hydrothermal waters have deposited chemicals that make some of the layers impervious.

The great heterogeneity in subsurface structure is reflected in the great variability in reservoir characteristics, especially temperature and salinity, from area to area as indicated in Table 10-5. For example, the temperature of Cerro Prieto is 350°C whereas that of nearby East Mesa is only 180°C.

Cerro Prieto, also in the Mexicali-Imperial Rift Valley, is at the southern end of the San Andreas fault system in a sedimentary alluvial-marine plain at the mouth of the Colorado River. It lies above a big granitic graben above which is 2500 m of sedimentary fill (Fig. 10-3). The graben and accompanying fill has resulted from spreading of the Gulf of California. Hot water and steam production is from interbedded sandstone and shales. At the nearby Pathé, the graben itself is the reservoir consisting of intensely fractured and altered andesites and basalts up to 1500 m thick. Productivity of the field is low, due to lack of lateral permeability in the volcanic materials.

Japan has geothermal environments basically similar to those of Kamchatka and New Zealand (Fig. 2-3). Most of the hot springs in Japan are closely associated with Quaternary volcanoes and domes of rhyolite and andesite rather than basalt. Some thermal fields, however, lie in highly faulted Tertiary volcanic

Table 10-5. Geothermal Energy Resources of Mexicali-Imperial Rift Valley (Adapted from White and Williams, 1975)

Location	Subsurface Temperature (°C)	Reservoir Volume (km³)	Heat Content (10^8 cal)	Mineral Content (ppm)
Cerro Prieto	350	—	*	22,000
Salton Sea	340	108	21	258,000
North Brawley	200	27	3	—
Heber	190	100	11	—
East Mesa	180	56	5.5	25,000
Border	160	1.8	0.2	—
East Brawley	135	3	0.2	—
Glamis	135	6	0.4	—
Dunes	135	9	0.6	—

* Producing 150 Mw of electricity.

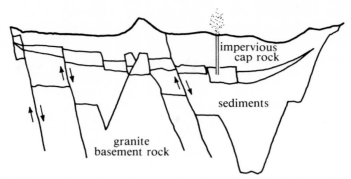

Figure 10-3. Geologic cross section of Cerro Prieto showing heat storage aquifer. (Adapted from Birsic, 1976.)

and granitic areas unrelated to Quaternary volcanism. The heat sources are both young volcanoes and deep intrusives of Tertiary and Late Mesozoic ages.

Relatively little surface thermal activity is present in the Matsukawa area, except for a wide area of hydrothermally altered rock. An approximately 1700 m thick sequence of porous, fractured welded tuff, and marine black shale and sandstone trap hot water and steam. These older volcanic reservoir rocks are capped by a 160 m thick layer of andesitic lavas from the Pleistocene volcano Marumori located about 1 km away.

The nearby Otake geothermal system is basically similar, except for more surface flow of hot water and steam. In both fields the boreholes take their flow from cracks and fissures rather than porous permeable media.

Hydrothermal activity at Pauzhetsk in southeastern Kamchatka (Fig. 2-3) is localized along fault margins of horsts within large regional depressions and at fault intersections of caldera collapse margins. Production is from Quaternary highly permeable pyroclastics interbedded with low permeability layers, all heavily normally faulted.

One resource base, considered to be immense, is under active exploitation near the Valle Grande caldera in northern New Mexico. It consists of hot dry rock, predominantly Precambrian banded granitic gneiss intruded by monzogranitic volcanic dikes which are devoid of fluids. Present activity is centered around an area lying immediately to the west of the caldera rim fracture zone. The basic development plan is to establish a geothermal reservoir by connecting two deeply drilled holes in geothermally heated impermeable rock through a system of fractures produced by pumping water into the holes under high pressure (Fig. 10-4). The first holes have been drilled to a depth of about 3000 m where the temperatures reach 173°C to 200°C. Heat is then extracted, after the hydraulic fracturing has been accomplished, by pumping cold water down into the deeper of the two holes. The numerous open fractures allow the water to come in direct contact with large surfaces of hot rock where it becomes heated. It is then brought to the surface up through the second hole under sufficient pressure to keep it from boiling and pumped through a heat exchanger. After losing its heat,

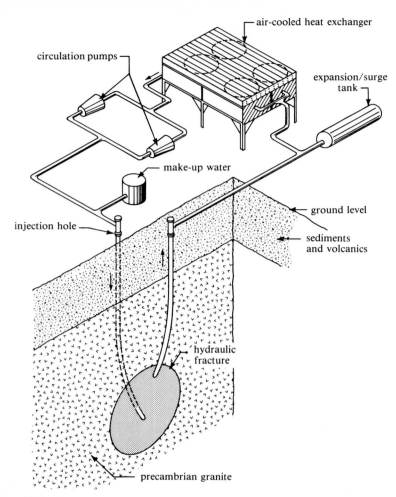

Figure 10-4. Conceptual diagram of hot dry rock geothermal experimental system. (After Tester et al., 1979.)

it is pumped back down the first hole, thus forming a completely self-contained, closed-loop recirculating system.

Potential sources of electric power are geopressurized waters located at great depth, 2000 m to 3500 m. These geopressurized areas occur in continuous belts extending for hundreds of kilometers, particularly along the Gulf Coast of the United States. The water stored in the deeply buried sand aquifers bounded by regional faults, was originally intercrystalline water residing in clays from which it has now been liberated by heat and pressure. The water is fresh and contains large quantities of dissolved oil and gases which are commercially extracted on a limited basis. Temperatures range from 150°C to 180°C; wellhead pressures

range from 25 to 40 megapascals; and production rates are likely to reach several million liters of water per day from each well.

The Carpathian geothermal area in Hungary is presumably a geopressurized region not associated with any obvious volcanic activity. It was discovered in the 1950s as a result of an extensive heat flow measurement program that found abnormally high heat flow, 2.0 to 3.4 μcal/s, about twice normal, over large areas. About 80 geothermal wells, 2000 m deep, have been drilled in the area and produce 80°C to 99°C water, at a shutdown wellhead pressure ranging from 0.3 to 0.5 megapascals. The basement rock is Paleozoic metamorphic and Mesozoic limestone, and dolomite formation. This is overlain with 6000 m of porous sedimentary rocks and sediments. The main hot water reservoir is in the Upper Pleiocine sandstone formation, consisting of sheets of sand and sandstone interlayered with siltstone. Water is taken from three to eight of these sandstone layers along a 100 to 300 m length of borehole.

10.3 Agricultural and Other Husbandry Uses

Perhaps the best-known use of geothermal fluids, aside from recreational and therapeutic purposes, is for agricultural purposes, particularly for greenhouses. The most famous of these are found in Iceland where tomatoes and bananas are harvested the year round in geothermally heated greenhouses. For this use the temperature of the geothermal fluid seldom exceeds 100°C and can be as low as 60°C. A fluid temperature of 40°C is quite adequate for the warming of soil outdoors and fish farming can utilize even lower temperatures.

One rather fanciful agricultural scheme that has been envisioned but not yet implemented highlights the great inherent possibilities. It combines greenhouse crops, mushroom culture, fish farming and biogas generation into an integrated biologic system with the waste or byproduct of each subsystem recovered and fed to the next cycle. The details are quite sensible and bear an examination. A family size module would occupy about two hectares and include the following:

- a geothermal source providing 2000 liters/min of fluid at a temperature of 120°C;
- two heat exchangers of 1750 kw capacity;
- three greenhouses, each covering 280 m^2 area;
- three mushroom houses, each with 370 m^2 area;
- six prawn ponds of 2500 m^2 each;
- six biogas digestion units, each with 65 m^2 daily biogas production capacity, yielding biogas and organic compost to be utilized as planting mulch and fertilizer for the agricultural activities and as a fertilizer to promote best growing conditions for the prawn ponds;
- two 15 kw engine-generator sets to be powered by biogas;
- absorption refrigerated temporary product storage area;

- a combination laboratory and office occupying 84 m^2; and
- circulation pumps and various appurtenant facilities.

Geothermal fluids are popular as the source of heat for greenhouses since even in temperate climates the cost of operating a greenhouse using fossil fuel usually amounts to 15 to 20 percent of the value of the products. In Iceland, radiator pipes are installed along the walls for heating. A water temperature of 80°C to 95°C is common. Where the water is hot enough it is used to sterilize the soil after each season. About 70 percent of the greenhouse area is used to grow tomatoes and cucumbers, the remaining 30 percent being used for roses, carnations, chrysanthemums, and potted plants. About 1000 tons of vegetables are grown in the 110,000 m^2 total area available in Iceland, saving some 20,000 metric tons of oil per year if the greenhouses were heated by fossil fuel.

Greenhouse activity has expanded tremendously in recent years in Hungary, the growing area now being about one square kilometer. Japan grows many of its flowers, fruits, and vegetables in geothermally heated greenhouses. New Zealand grows many of its mushrooms by using the geothermal fluids directly without the use of heat exchangers, the source water being hot enough to sterilize the soil when necessary.

Russia has numerous greenhouse centers that produce important quantities of vegetables. A single unit generally covers about 1000 m^2 and produces 15 to 20 kg of tomatoes and cucumbers per year per square meter. To accomplish the same production with fossil fuel would take about a metric ton of coal.

Often the greenhouses in Russia are surrounded by outside areas where the soil is warmed with geothermal fluids. This greatly extends the growing season. Similar soil-warming experiments carried out in the United States increased the yield of corn silage by 45 percent, of tomatoes by 50 percent, of soybean silage by 65 percent, and of bush beans by 39 percent, plus improving the quality of the crops.

Geothermal fluids also find wide application in animal husbandry. In Hungary, they heat cattle stalls, milking rooms, pigsties, and chicken houses; in Japan, they aid in hatching eggs and raising poultry; and in New Zealand, they help to biograde the wastes from pigsties which proceeds more rapidly at higher temperatures.

Some other agricultural uses include washing and drying of wool, production of dried milk, casein, and sucrose, and crop drying.

The breeding of fish, especially in Iceland and Japan, relies heavily on geothermal water. The Kallafjord Experimental Fish Farm, the largest in Iceland, keeps the water for hatching salmon at a temperature of 10°C to 14°C using indirect heating by geothermal water. This hatching produces about 300,000 salmon smelts per year which either migrate to the sea or are released into rivers. Another smelt-breeding station uses warm spring water directly. In Japan eels are bred for food in a 23°C mixture of river and geothermal water. Alligators and crocodiles are bred as a tourist attraction in 28°C to 32°C geothermal waste waters.

10.4 Space Heating and Cooling

Iceland was the first country to utilize geothermal fluids for large scale space heating. At present almost 60 percent of its population is served by fields at Reykjavik, Svartsengi, Krisuvik, and Hengill in the southwest, and Namafjall, Krafla, and Theistareyki in the north. The Reykjavik system is the largest in the world. Iceland is ideally suited for such a project by virtue of its cold, yet temperate climate which requires almost year-round home heating, its complete lack of significant indigenous fossil fuel sources, and its abundance of geothermal resources.

After early pilot efforts, extensive drilling, 15 to 20 km from Reykjavik for hot water in the early 1930s yielded enough to heat 2300 homes housing 30,000 people, and all of the public buildings. This original system has been operating continuously since 1943. Another field was opened up 3 km of Reykir in 1950. All together 72 boreholes in these two spring areas supplied Reykjavik, producing a total of 330 liters/s of 86°C water. The boreholes vary in depth from about 300 m to 770 m.

More recently, two additional fields have been developed within the city limits. One yields 300 liters/s of 128°C water from relatively deep holes, 700 to 2200 m deep, and the other 165 liters/s of 105°C water. This newer source of water, averaging 119°C and totalling 475 liters/s flows from wells that are equipped with pumps placed down within at depths of 110 m to 120 m.

A schematic drawing of the Reykjavik system is shown in Fig. 10-5. The 86°C water yielded by the Reykir wells flows under gravity into the collecting tanks (1) from which it is pumped through two collocated 15.3 kg long, 35 cm diameter pipes to 8000 m³ capacity storage tanks (4) located on a hill within Reykjavik. A standby oil-fired heating plant (2) is used on cold days to raise the temperature of the water which is normally maintained at 90° C in the storage tanks.

Most of the homes, now numbering about 8700 and housing 72,000 inhabitants, are connected to this technically and economically highly successful system. The houses are supplied with 80°C water both for heating and domestic use. Reykir water is simply pumped from the storage tanks to the houses through a single pipe system (8) with some being dumped in the sewer and the remainder mixed with the hotter Reykjavik water to lower the latter's temperature. As the population has increased, several substations have been built for mixing and distributing. The entire system except for the deep well pumps operates automatically. Development in other parts of Iceland has been slower.

At Klamath Falls, Oregon, hot, 130°C to 230°C water obtained from 400, 30 m to 600m deep, wells is used to heat approximately 500 structures. Some of the water is used directly from the wells but the bulk of the heat is obtained through the use of down-hole, hairpin heat exchangers using water drawn from the regular city mains as the circulating fluid. About 170 homes in Boise, Idaho, are directly heated by 59°C water discharged from two wells. Many areas are under

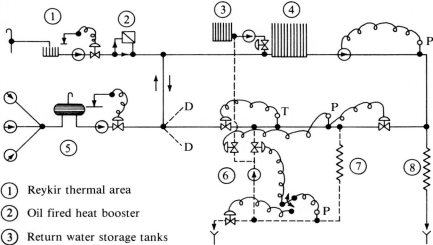

① Reykir thermal area

② Oil fired heat booster

③ Return water storage tanks

④ Storage tanks (supply water)

⑤ Reykjavik thermal area

⑥ District pumping station

⑦ Two pipe distribution system

⑧ Single pipe distribution system

P   Pressure control
T   Temperature control
D   District substation

Figure 10-5. Schematic diagram of space heating system installed in Reykjavik. (Adapted from Einarsson, 1973.)

development, especially in the Basin and Range Province of the western United States.

The 20,000 inhabitants of Rotorua make extensive use of their locally available geothermal resource. It is most inefficient as individuals have been allowed to drill more than 1000 boreholes within the city limits, often right next to their residences. The numerous boreholes, issuing a mixture of water and steam are connected to single houses, small groups of houses, public building complexes, hotels, and hospitals. The most outstanding project has been the air-conditioning system at the Rotorua International Hotel which both cools and heats depending upon the weather. It is designed for extreme ambient temperatures ranging from −4°C to 30°C. The maximum heating load is 0.5 Gcal/hr (2,000,000 BTU). The 130 ton, 0.39 Gcal/hr output unit requires a heat input of 0.575 Gcal/hr, about 50 percent more than its output. A well producing 150°C water at a pressure of 0.6 megapascals supplies the heat energy. A heat exchanger transfers the heat of the

geothermal fluid to fresh water contained in a closed circuit, heating it to a temperature of 120°C, which then supplies heat to the radiators, domestic water heaters, and the air-conditioning unit.

The Russians have devoted a great deal of attention to the development of technically and economically sound geothermal space heating systems, some of these involving the combination of different types of energy sources. One of their earlier installations at Makhach-Kala is fed by a deep well originally drilled for oil. It produces 23 liters/s of 63°C hot water which is used to supply dozens of dwellings and industrial buildings. Additional wells supply 70 liters/s to 15,000 inhabitants. Some other Russian installations and uses of geothermal waters are listed in Table 10-6.

Very extensive but relatively low temperature geothermal resources were discovered in Hungary in the late 1950s while drilling for oil and gas and preliminary plans were formulated for utilizing them for domestic heating. Subsequent discovery of rich and oil and gas deposits precluded the use of geothermal fluids that are now being used mainly for heating greenhouses and for animal husbandry, a greatly expanding activity.

The homes in several Japanese cities are now heated geothermically. Among

Table 10-6. Some USSR Space Heating Installations (Adapted from Einarsson, 1973)

Location	Description	Energy Consumption (Gcal/hr)
Astarinsk district (Azerbaidjan)	15 hectares greenhouses	46.5
Zgondidi town (Georgia)	Heating of flats, public buildings, greenhouses, swimming pools	50.0
Iserbach town (Daghestan)	District heating for 7500 inhabitants, industrial uses	6.0
Caspillsk town (Daghestan)	District heating for 5000 inhabitants, hot water supply	5.0
Massalinski district	15 hectares greenhouses	46.5
Mendji (Georgia)	Heating of meterological station, agricultural uses	2.0
Paratounka (Kamchatka)	Heating of 3 apartment houses with 48 units each	0.55
Ternahir (Kamchatka)	5 hectares greenhouses, 5 hectares soil heating	19.5
Zaichi (Georgia)	Heating of meteorological station, greenhouses, baths	2.1
Cherkest (Stravropol)	District heating for 18,200 inhabitants, industrial uses, greenhouses	2.2

the projects are an 11.5 km long transmission line carrying 14 liters/s of 70°C water from Sarujura springs to the town of Towata; a 12 km long line in the Okawa area which supplies 3000 houses spread over a 260 hectare area with 22 liters/s of 70°C water; and a heating system comprised of 140 houses and 34 hotels at Aomori supplied from the Asamushi hot springs at a flow rate of 22 liters/s and 60°C water.

10.5 Industrial Processing

Practically the entire range of temperatures of available geothermal fluids, steam, hot water, and mixtures of them can be used one way or another in industrial processing (Table 10-1). In general, industrial applications require the highest temperatures usually in the form of steam, whereas space heating and many agricultural projects can be accomplished with relatively low temperature water.

Among the industrial uses of geothermal energy are process heating such as for drying and distillation, refrigeration, de-icing, and tempering in various mining and material handling operations. The fluids are also the source of many salts and other valuable chemicals.

Hot geothermal energy for cooling for industrial purposes is made possible through the use of lithium bromide absorption equipment. This system is found at the food processing plant at Brady's Hot Springs, Nevada, and the fish freezing facility at Akureyri, Iceland.

The pulp and paper mills at Kawerau, New Zealand, were one of the first major industrial plants to use steam extensively for processing, and indeed, the site of the mills was selected with that in mind, close to a large source of geothermal energy along the Tarawera River. More than 180,000 kg/hr of natural steam are expended in the production of newsprint, pulp, and sawn timber. Pulp production by the Kraft process takes the major portion of the steam although the conversion of the pulp into finished paper requires considerable steam especially for drying. The Kraft process consists of using steam in the presence of chemicals to digest wood chips, separating the cellulose from the black liquor so formed, and finally washing and drying the cellulose to form the pulp. The black liquor is usually evaporated to recover the chemicals it contains.

Figure 10-6 shows the layout of the wells, steam lines, and plant units. The geothermal fluid is flashed at the wellhead to produce wet steam. The steam is separated from the water and transmitted by pipelines. One, 30 cm in diameter, flows 36,000 kg/hr of steam at 1.4 megapascals; the other, a 60 cm diameter pipe transmits 145,000 kg/hr of steam at 0.7 megapascals. This high-pressure steam is first used directly for timber drying and powering log-handling equipment after which it is run through a heat exchanger to produce the 1.0 megapascal clean steam used in the pulp processing. The resultant low-pressure natural steam is used further in a heat exchanger to generate large quantities of 0.4 megapascal steam, some of which drives a 10 Mw noncondensing electric generator, the

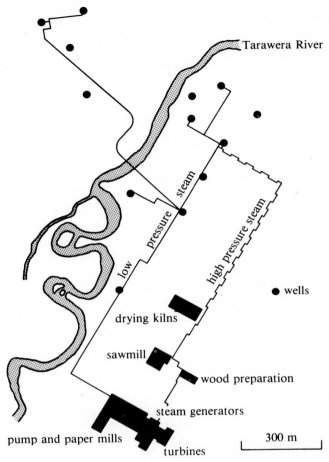

Figure 10-6. Pulp and paper mills at Kawerau, New Zealand. (Adapted from Lindal, 1973.)

main source of electrical power in the plant, and the remainder evaporates the black liquor.

Some of the newest and most promising applications of geothermal energy are in the fields of mining and upgrading of minerals. Its utilization in the profitable recovery of diatomaceous earth at Lake Myvatn in northern Iceland is an outstanding example of where a readily available cheap heat source made a hitherto worthless mineral deposit economically attractive. Normally only dry deposits of the material can be mined profitably. The huge deposits of diatomaceous earth at Lake Myvatn were of high quality but extremely wet, the very porous diatomaceous earth tending to retain three to four times its own weight of water even after filtration. The recovery plant was built only 600 m from the Namafjall geothermal area where large quantities of steam were available. The diatomace-

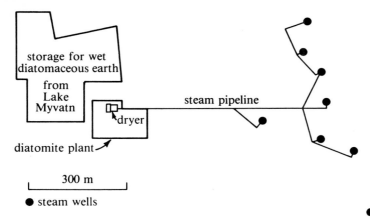

Figure 10-7. Diatomaceous earth drying plant at Myvatn, Iceland, showing steam supply system. (Adapted from Lindal, 1973.)

ous earth is dredged from the bottom of the lake, 3 km away, and pumped as a slurry to the plant where 50 metric tons per hour of 1.0 megapascal, 180°C hot water is available from nearby wells. A schematic of the plant is shown in Fig. 10-7.

10.6 Electric Power Production

The largest practical application, approximately 33 percent, of geothermal energy is for the generation of electrical power. Since 1950, the technology has progressed rapidly, with now more than 1000 Mw worldwide on line. The process is inefficient by fossil fuel standards because of the low temperature of the geothermal fluids, 230°C to 300°C. Nevertheless, the conversion of geothermal energy into electricity makes possible the utilization of its energy over long distances.

On a nationwide scale, experts believe that there is geologic potential to produce enough geothermal energy to supply 20,000 Mw of generating capacity by the year 2000, equivalent to 700,000 barrels of oil a day or 8.7 percent of USA crude oil production. Engineers estimate that someday more than 10 percent of all electric power used in the USA will be geothermically generated. Practical utilization of geothermal energy offers a gigantic potential. The USSR, for example, estimates that their geothermal potential is probably equal to their reserves of petroleum.

Presently operating electrical generating systems listed in Table 10-7, fall into the following categories as regards the form of geothermal fluid used:

● Dry steam, steam directly from the reservoir, after minor cleanup used as a working fluid for a low pressure steam turbine;

Table 10-7. Geothermal Power Plants

Plants	Mw Capacity	Initial Operation
Dry Steam		
Italy		
Larderello	380	1904
Monte Amiata	26	1967
USA		
The Geysers, CA	502	1960
Japan		
Matsukawa	22	1966
Onikobe	25	1975
Flashed Steam		
New Zealand		
Wairakei	192	1958
Kawerau	10	1969
Japan		
Otake	23	1967
Hatchabaru	50	1976
Katsukonda	50	1977
Mexico		
Pathé	4	1958
Cerro Prieto	70(?)	1973
Iceland		
Kafla	70	1977
Philippines		
Tiwi	10	1969
USSR		
Pauzhetsk	6	1967
El Salvador		
Ahuachapan	30	1975
Binary cycle		
USSR		
Paratunka	1	1967
USA		
Imperial Valley, CA	10–50	Late 1970s to 1980

- Flashed steam, hot water flashed by dropping its pressure to form steam and water that are separated; the water is rejected and the steam is used to drive a low pressure turbine; and
- Binary cycle, heat from hot water transferred to a secondary working fluid such as freon or isobutane which is vaporized and used to drive a turbine.

Regardless of the system used, the available work and the conversion efficiency both increase with increase of the temperature of the geothermal fluid. The increase in available work is particularly marked (Figs. 10-8 and 10-9).

The oldest, since 1904, and until recently the largest generating complex is

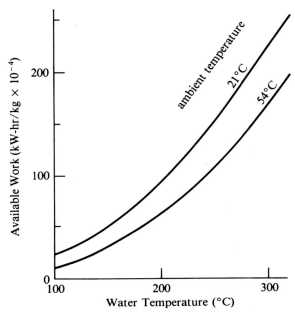

Figure 10-8. Available work, the maximum reversible work, for water at its saturation pressure and temperature discharging to ambient conditions of 0.1 megapascals pressure at the temperatures shown. (From Wahl, 1977.)

located at Larderello, Italy, where 13 generating plants produce 365 Mw of electricity. An additional 40 Mw are generated at the nearby Monte Amiata and Travale fields. The electric railroads in Italy are run almost exclusively by power generated at the steam fields of Larderello. ⊣ᴦᴀᴉᴺᴏ

The Geysers project is the largest geothermal project in the world. Both Larderello and The Geysers are examples of dry steam fields where the turbines are being driven with steam essentially as it comes from the wells. This development began in 1957, and by 1960 it was producing enough steam to power a 12.5 Mw generating plant. Today enough steam is produced to power a generating capacity exceeding 660 Mw, which is sufficient to supply all the electrical requirements of a typical city of over 600,000. Currently, an additional 330 Mw of generating capacity is under construction to be on line by 1980. This will bring the total capability of the field to over 900 Mw, enough to supply about one-third the needs of the San Francisco Bay area.

The fact that this particular resource can be used so directly and efficiently has been a key element in the successful development of The Geysers to its current size. Development has been in increments of 55 to 110 Mw which appear to be optimum blocks of power for the number of wells required, the pipeline distance, and the size and cost of the turbines.

The steam gathering layout of The Geysers is shown in Fig. 10-10. Production is from graywacke sandstone in which it is very difficult to drill. The reservoir

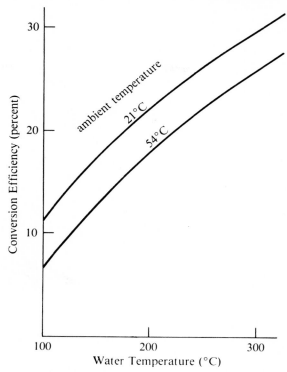

Figure 10-9. Conversion efficiency as a function of temperature under the same conditions as Fig. 10-8. (From Wahl, 1977.)

pressure is about 3 megapascals at a temperature of 205°C. The wellhead pressure is 0.8 megapascals with the turbines operating at 0.5 to 0.7 megapascals. Each individual generating unit consists of two collocated 55 Mw generating plants (Nos. 7 and 8 in the figure). The steam available is 70,000 kg/hr/well. Each 55 Mw unit needs seven wells (490,000 kg/hr) to supply it. The wells are 600 to 3000 m deep and spaced at one well per 2 hectares.

The most economical operation is obtained using dry 179°C steam entering the turbine through a 60 cm diameter line at a pressure of 0.7 megapascals and exhausting at 10 cm of Hg absolute out of 58 cm diameter last-stage turbine blades. Determining the optimum pressure is tricky. At higher pressure, the increased energy available from the steam is more than offset by the diminished steam production rates from the wells. At lower pressure, the increased production rate from the wells is offset by the decreased available energy in the steam.

Loss of steam head is due to pressure decrease in reservoir, not to scale building up in the pipes. To prevent corrosion, the main dry steam distributing pipes, valves, and strainers are made of carbon steel; stainless steel is used for most parts that become wetted. Generally, pipes run from 30 to 70 cm in diameter, increasing in size as steam is gathered from more and more wells. Surface

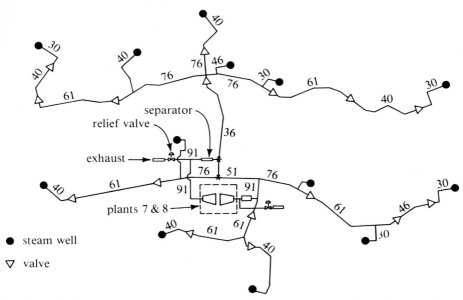

Figure 10-10. Steam gathering system at The Geysers. Numbers on steam lines are pipe diameters in cm. (After Budd, 1975.)

piping must be large enough to carry steam but not so large as to lower steam pressure before it reaches the turbine.

Near each wellhead is a vessel for the separation of liquid and solid particulate matter from the steam. Centrifugal type horizontal separators, 99 percent effective for particles greater than 10 microns in size are used.

Eight percent of the exhaust steam is evaporated in cooling towers of conventional design. Approximately one percent of the vapor which is noncondensible is discharged as a gas out of the stack with a percentage composition of:

Carbon dioxide	0.79
Hydrogen sulfide	0.05
Ammonia	0.07
Nitrogen	0.03
Hydrogen	0.01
Methane	trace

The remaining condensed waste water, which contains boron, ammonia, bicarbonate, sulfate, and finely divided sulfur, is reinjected into the ground in old wells.

At Wairakei, about 100 wells supply 1,200,000 kg/hr of steam, flashed from hot water, to generate about 190 Mw of electricity. Power is producd by two 6.5 Mw, eight 11.2 Mw, and three 3.0 Mw generators driven by steam turbines. The steam from about half the wells is at 0.9 megapascals and the rest at 0.34

megapascals. Only about 20 percent by weight of the hot water flowing from the wells flashes into the steam. Additional energy is extracted from some of the newer wells by flashing the water from the wellhead separator a second time at reduced pressure to produce low pressure steam and occasionally flashed a third time at still lower pressure. About 1800 liters/s of hot water from the final stages of the separators flows through concrete channels to the Waikato River.

The steam moves in pipelines from the steam field to the power station 2.5 km away. The several lines total about 25 km in length and range in diameter from 48 cm to 107 cm. Loops every 300 m allow for thermal expansion; in the main line they are 5 m high to take care of 6 m of expansion. A 107 cm diameter, 1.6 km long pipe, the largest of its kind in the world, transmits 270,000 kg/hr of low pressure steam to the station. Water is pumped from the Waikato River to condense the exhausted steam from the turbines. Both the condensate and this circulating cooling water are returned to the river.

At Cerro Prieto (Fig. 10-11) steam flashed from the extraordinarily hot, 340°C water drives the two 75 Mw electric generators. The water is flashed twice; first, at the wellhead where centrifugal separators remove up to 99.9 percent of the water. The separated water is then flashed the second time at a lower pressure in a different part of the turbine. Standard multistage turbines are used giving an overall efficiency of only 11 percent, but still economically viable.

Figure 10-11. Cerro Prieto electric generating plant in early stages of construction.

A 30 Mw plant has just been completed at Ahuachapan, El Salvador; and Guadelupe, Chile, and Nicaragua all have projects underway.

About 60 Mw are being generated by a recently completed plant at Namafjall.

The principal geothermal plants in Japan produce 22 Mw at Matsukawa, 25 Mw at Onikobe, 13 Mw at Otaki, 10 Mw at Onomi, and 5.5 Mw at Hatchoburu.

Binary systems have been designed to utilize highly mineralized or relatively cool geothermal fluids by extracting the contained heat in them without rendering the equipment inoperable. The operation, effective for both quite hot fluids as well as those as cool as 38°C, involves the use of low-boiling-point gas, freon or isobutane, in much the same way as it is used in refrigeration. The heat vaporizes the fluid and the evolved gas is used to power turbines that drive generators. With this scheme, the heat from otherwise unusable fluids can be salvaged.

Two freon pilot plants have been operating in the USSR since the mid 1960s. Water at a temperature of 91.5°C is used in a 0.34 Mw plant; and at 180°C in a 0.75 Mw one.

One experimental pilot plant incorporating a binary system for power production is also located in the Salton Sea area. The system flashes the high salinity brine into steam through separators. The steam thus formed then flows through scrubbers and into heat exchangers where it heats isobutane sealed in a closed system. On being heated the isobutane vaporizes and the vapor is used as the working fluid to drive a turbogenerator. The overall thermal efficiency of this system is about 15 percent. The isobutane is exhausted as a liquid from the turbine and returned to the heat exchanger where it is again vaporized. The steam is also recycled. When it leaves the heat exchanger, the condensate is combined with the original brines and reinjected into the reservoir through another well (Fig. 10-12).

The first application of the Valle Grande resource will be to operate a small generating plant utilizing isobutane in a binary system.

10.7 Economic and Environmental Aspects

The development of a geothermal field is based on an estimate of its energy potential, together with estimates of the economics of development. Unfortunately projections cannot be made with the same degree of precision as can be done for oil and gas fields. Certainly not all geothermal fields are amenable to economic exploitation, and in order to determine whether a field can be profitably put to use, it is necessary to expend fairly formidable sums in carrying out exploration. If the results are positive, then exploration costs will have been well spent. Where available, geothermal energy can provide one of the cheapest sources of heat in the world. Even if the exploitation of a geothermal energy resource requires fairly heavy initial capital expenditures for exploration and development, the subsequent operating costs are relatively low. In general, development has been very slow, largely because in most instances other energy resources are cheaper, better understood, and easier to evaluate. As the fossil fuel

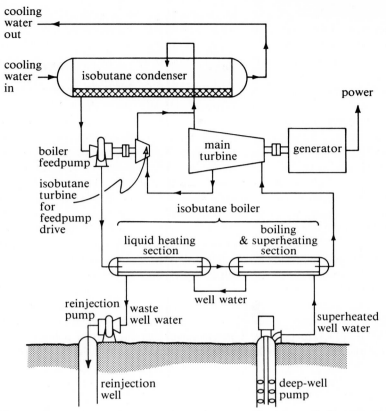

Figure 10-12. Binary system utilizing isobutane which derives its heat from geothermal steam. (From Anderson, 1973.)

supply is depleted, some of these considerations will become less important, especially the cost factor.

The usual capital costs are similar to those in developing an oil field. First there is the exploration which includes topographic and geologic surveys, geophysical and geochemical investigations, exploratory drilling, and field investigations. Drilling, the sinking of production wells, including dry holes, at locations which appear to have the greatest probability of achieving high production as determined by exploration is very expensive. Wellhead equipment must be supplied and installed: suitable valving, separators, silencers, and instrumentation, with incidental integral pipework. Collection pipework must also be supplied and installed: a suitable system, with lagging, drainage, and expansion facilities for the purpose of collecting the geothermal fluids from the wellheads and delivering them to the point of need.

The cost of geothermal energy increases sharply with increasing distance it has to be transmitted. Only when the plant can be situated essentially adjacent to

the geothermal field can the low cost of steam and hot water, the principal advantage of geothermal energy, be realized. The cost of natural steam will vary from place to place but as a rule of thumb for a reasonably favorable place, the cost of primary steam ranges from one-tenth to one-fifth of that produced by fossil fuel.

Recurring costs, essentially independent of the quantity of heat delivered to the point of use, are: interest on capital expenditures; depreciation of capital assets or amortization of loan; maintenance and repairs of pipework, valves, wellhead equipment, and well; drilling of new wells from time to time to replace depleted ones; and salaries and wages of the operating, inspecting, and supervising staff.

One of the biggest unknowns is the lifetime of the resource. Steam production from individual wells has been decreasing with time at both The Geysers and Larderello (Fig. 10-13). There is also some indication that total productivity of the fields may be decreasing somewhat. At Wairakei, the rate of evolution of hot water and steam has decreased considerably during the 20 years of the operation. The exact location of the source and extent of the water residing in the basins at Wairakei is obscure, although geologists seem to feel that the supply will probably last until near the end of this century.

The economic feasibility of using geothermal energy depends on whether it can compete economically with other available sources of energy, fossil fueled, nuclear, and solar. Some 50 countries, especially developing countries, now active or interested in geothermal exploration, are attracted by the relatively

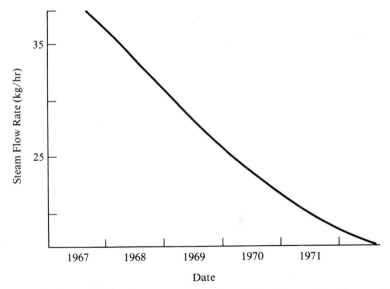

Figure 10-13. Steam production rate as a function of time in operation for geothermal steam wells at The Geysers. (Adapted from Budd, 1975.)

small capacity of the geothermal power units, 100 Mw, as opposed to the larger units, 1000 Mw, required to use fossil fueled power economically.

Table 10-8 indicates the relative economic values of various segments of the geothermal industry worldwide as of 1975.

Several environmental problems have developed in connection with the exploitation of geothermal resources. Pollution problems can be atmospheric from the disposal of gas, thermal from the waste heat, and chemical from the dissolved salts in the waste water. The seriousness of any one problem varies considerably from field to field.

At many fields, atmospheric pollution is not a serious problem since the amount of noncondensible gas discharge is very small and can simply be discharged through stacks. At others, discharge of hydrogen sulfide gas into the atmosphere near power generating facilities greatly deteriorates air quality. At The Geysers, where the hydrogen sulfide concentration in the geothermal steam is 150 to 300 ppm by weight, emission of the sulfide increased from 14 kg/hr in 1960 when electrical production was very low to 750 kg/hr in 1975 when production had reached about 400 Mw. Although the high level did not present a serious health problem, it provoked many nearby residents to complain about the pervasive odor of rotten eggs. It is interesting to note that the discharge of hydrogen sulfide is about the same amount as would be given off by a fossil fuel plant of the same size.

As a consequence, The Geysers is now incorporating in its design, devices for removing hydrogen sulfide and other obnoxious gases. In trial tests, upstream scrubbers using a copper sulfate solution have been able to remove over 90 percent of the hydrogen sulfide from the steam going into the generators as well as about 80 percent of the ammonia gas and hydrogen borate. Iron used as a catalyst in the system also shows much promise. Controlled amounts are added to the circulating cooling water to destroy the hydrogen sulfide by oxidizing dissolved hydrogen sulfide to elemental sulfur.

Thermal pollution is a problem at Wairakei where a local river is used for the disposal of waste water. Thermal pollution can affect fish life and encourage undesirable plant growth. In general, a facility disposes of its waste water either by reinjection into the reservoir, by discharge into surface waters, or by lagooning and ponding. Table 10-9 lists practices at several fields under exploitation.

Table 10-8. Relative Economic Values of Segments of Geothermal Industry (Adapted from Peterson and El-Ramly, 1976)

Market Segment	Percent of Market
Balneology/Tourism	50.8
Electrical	27.6
Agricultural	12.7
Space Heating	7.5
Industrial	1.4

Table 10-9. Practices for Disposal of Waste Water (Adapted from Morrison et al., 1977)

Facility	Practice
The Geysers, USA	Reinjection
Wairakei, NZ	Discharge to Waikato River (60°C)
Cerro Prieto, Mexico	Discharge to evaporation and sedimentation ponds
Imperial Valley, USA	Reinjection and ponding
Ahuachapan, El Salvador	Reinjection
Tiwi, Philippines	Surface disposal

The problems associated with chemical pollution can be serious enough to prevent or hinder the growth of an otherwise promising field. Steam fields are usually not troubled in this way. One common solution is reinjection of the waste water into the ground either through old, nonproducing wells or specially drilled ones. At Wairakei, although the chemical pollution problems is not very serious, silica deposits must be removed annually from the channels that carry away the waste water from the steam separators. The silica problem at Otake is more serious. Until a remedy, primarily suitable agitation, was found, the waste water pipelines tended to clog up completely.

The environmental impact of the development and utilization of a geothermal resource will vary widely with the quality of the steam and water that emerge and will be dependent on whether subterranean pressure prevents returning them to earth. If the steam and water must be disposed surficially, the problem may become formidable, although adequate water for cooling is an asset and relatively pure water may find agricultural and other uses as at Raft River, Idaho.

In extensive sedimentary regions such as the Imperial Valley, extraction of fluids could easily lead to subsidence of the land to such an extent that flow through irrigation ditches is affected. Both vertical and horizontal ground movements have been substantial at Wairakei where a roughly elliptical, 1200 m diameter, 3 to 6 m deep, dish-shaped depression has formed within a few thousand meters of several of the producing wells. The maximum rate of subsidence has been about 40 cm per year. The movement appears to be caused by the collapse of the mudstone cap rock when the pressure supporting it decreases as large amounts of fluids are withdrawn from underneath it. Ground movement is more of a nuisance than a catastrophe, but it requires considerable field maintenance.

In most respects, the hot dry rock system, still in experimental stages, seems the most environmentally satisfactory source of geothermal energy, its main disadvantage being the large amounts of surface water required. An original fear that the hydrofracturing would produce earthquakes or other geologic instabilities has lessened as more has been learned about earthquake initiation.

Noise pollution may be serious in populated areas, although not insurmountable. Steam escaping under high pressure is ear-splitting.

Heat rejection is another significant local environmental effect. If the ten 1000 Mw geothermal generating plants are installed as contemplated in Imperial Valley, the total rejected heat to the 2500 km^2 area that they would occupy would equal 5 percent of the solar energy reaching that area. Such an increment of heat could easily influence significantly the local weather pattern.

One objection voiced by environmental groups is the modification of the existing terrain by the pipelines and power plants of a producing geothermal field. The Geysers and Larderello are not particularly out of harmony with their surroundings and at Larderello, the land amidst the installation is being used for agricultural purposes. The chief impact comes during the construction phase but is confined to the area of the field and causes nothing like the disruption produced by strip mining coal.

Geothermally generated electricity, as with other forms of onsite generation, must be carried to where it is needed. This means an expansion of transmission lines, with their attendant problems of land use, visual pollution, and energy loss. None of these problems are confined to geothermal energy production. And in our energy-hungry and wasteful world, a balance must be drawn between the interests of the environmentalists, the producers, and the ultimate users. Fortunately, there are still places in the world where one can retreat to enjoy the aesthetics of geothermal energy—the geyser fields.

Appendix

Geologic Time (Stratigraphic Column)

Era	Period	Epoch	
Cenozoic	Quaternary	Holocene (Recent) Pleistocene	3 million years
	Tertiary	Pliocene Miocene Oligocene Eocene Paleocene	70 million years
Mesozoic	Cretaceous Jurassic Triassic		225–230 million years
Paleozoic	Permian Carboniferous { Pennsylvanian Mississippian Devonian		350 million years
	Silurian Ordovician Cambrian		500 million years
Precambrian			600–3500 million years

Chapter References

Chapter 1. Geysers of the World

Allen and Day, 1935; Barth, 1950; Birch and Kennedy, 1973; Brannock, Fix, Gianalla, and White, 1948; Bryan, 1979; Bunsen, 1848; Dana, 1894; Einarsson, 1967; Gippenreiter, 1969; Golovina and Malov, 1960; Grew, 1925; Hague, 1904; Healy, 1953; Hochstetter, 1867; Honda and Terada, 1906; Iwasaki, 1962; Keefer, 1972; Kimura, 1949; Krug von Nidda, 1837; Lloyd, 1965, 1975; Lloyd and Keam, 1974; Mackenzie, 1842; Munby, 1902; Murbarger, 1956; Noguchi, 1956; Nolan and Anderson, 1934; Peale, 1883; Rinehart, 1975b; Steinberg, Merzhanov, and Steinberg, 1978; Waring, 1965; White, 1967.

Chapter 2. The Geologic, Thermal, and Hydrologic State of the Earth

Allen and Day, 1935; Anderson, Anderegg, and Lawler, 1978; Bullard, 1973; Bunsen, 1848; Carr and Quinlivan, 1966; Dawson, 1964; Eaton, Christiansen, Iyer, Pitt, Mabey, Blank, Jr., Zietz, and Gettings, 1975; Fenner, 1936; Healy, 1953; Hochstetter, 1867; Iwasaki, 1962; Keefer, 1972; Kimura, 1949; Lloyd, 1975; Rinehart, 1960a, 1970a; Stacey, 1969; Steinberg, Merzhanov, and Steinberg, 1978; Thompson and Burke, 1974; Thorkelsson, 1940; Truesdell and Fournier, 1975; Verhoogen, Turner, Weiss, Wahrhaftig, and Fyfe, 1970; Wahl, 1977; Waring, 1965; White, 1965, 1967; White, Fournier, Muffler, and Truesdell, 1975.

Chapter 3. Fundamentals of Geyser Operation

Anderson, Anderegg, and Lawler, 1978; Barth, 1950; Birch and Kennedy, 1973; Bloss and Barth, 1949; Bryan, 1979; Bunsen, 1848; Dench, 1973; Einarsson,

1967; Golovina and Malov, 1960; Honda and Terada, 1906; Lloyd, 1975; Nekhoroshev, 1959; Noguchi, 1956; Normura, 1954; Thorkelsson, 1928, 1940; Waring, 1965; White, 1967.

Chapter 4. The Role of Gases in Geysers

Allen and Day, 1935; Barth, 1950; Bloss and Barth, 1949; Brannock, Fix, Gianalla, and White, 1948; Buchtala, 1928; Dana, 1894; Iwasaki, 1962; Mal'tsev and Bubhol'ts, 1961; Thorkelsson, 1910, 1928, 1940; Waring, 1965; White, 1967.

Chapter 5. Chemistry of Geothermal Waters

Allen and Day, 1935; Barth, 1950; Brannock, Fix, Gianalla, and White, 1948; Bullard, 1973; Bunsen, 1848; Fournier and Rowe, 1966; Healy, 1953; Hochstetter, 1867; Iwasaki, 1962; Lloyd, 1975; Rowe, Fournier, and Morey, 1973; Sigvaldason and White, 1973; Truesdell and Fournier, 1975, 1977; Tyndall, 1895; Waring, 1965; White, 1967.

Chapter 6. Geyser Area Complexes

Allen and Day, 1935; Anderson, Anderegg, and Lawler, 1978; Barth, 1950; Benseman, 1965; Birch and Kennedy, 1972; Bloss and Barth, 1949; Bunsen, 1848; Hague, 1904; Hochstetter, 1867; Iwasaki, 1962; Keefer, 1972; Lloyd, 1975; Marler, 1951; Nolan and Anderson, 1934; Rinehart, 1969a, 1970b; Rowe, Fournier, and Morey, 1965; Stearns, 1933; Thorkelsson, 1910, 1940; White, 1967.

Chapter 7. Environmental Aspects of Geysers

Allen and Day, 1935; Anderson, Anderegg, and Lawler, 1978; Brock, 1978; Bunsen, 1848; Dana, 1894; Einarsson, 1967; Fischer, 1966; Krug von Nidda, 1837; Nicholls and Rinehart, 1967; Nogoshi and Motoya, 1962; Rinehart, 1965, 1968a, 1972, 1974; Thorkelsson, 1910, 1940; Tyndall, 1895.

Chapter 8. Temporal Changes in Geyser Activity and Their Causes

Allen and Day, 1935; Barth, 1950; Einarsson, 1967; Honda and Terada, 1906; Iwasaki, 1962; Keefer, 1972; Krug von Nidda, 1837; Lloyd, 1975; Marler, 1954, 1964; Marler and White, 1977; Nomura, 1954; Rinehart, 1972a, 1972b, 1974, 1976b; Thorkelsson, 1910, 1940; White, 1967.

Chapter 9. Man's Influence on Geyser Activity

Allen and Day, 1935; Barth, 1950; Birch and Kennedy, 1973; Grew, 1925; Grindley, 1970; Hague, 1904; Iwasaki, 1962; Keefer, 1972; Krug von Nidda, 1837; Lloyd, 1975; Noguchi, 1956; Nomura, 1954; Steinberg, Merzhanov, and Steinberg, 1978; Thorkelsson, 1940; White, 1967.

Chapter 10. Practical Uses of Geothermal Fluids

Anderson, 1973; Banwell, 1973; Budd, 1975; Dench, 1973; Einarsson, 1973; Hammond 1972; Kruger and Otte, 1973; Lindal, 1973; Milora and Tester, 1976; Morrison, Saint, and Weaver, 1977; Otte, 1979; Peterson and El-Ramly, 1976; Rinehart, 1975, 1976a; Tester, Torris, Cummings, and Bivins, 1979; Wahl, 1977; White, 1965, 1967; White and Williams, 1975.

Bibliography

Allen, E. T. and A. L. Day, 1935. *Hot Springs of the Yellowstone National Park, Publ 466*. 525 pp. Carnegie Institution of Washington, Washington, D.C.

Anderson, J. H., 1973. The vapor-turbine cycle for geothermal power generation. In *Geothermal Energy*. 163–176. P. Kruger and C. Otte, eds. Stanford University Press, Stanford.

Anderson, L. W., J. W. Anderegg, and J. E. Lawler, 1978. Model geysers. *Amer. J. Sci.* **278,** 725–738.

Armstead, H. C. H., ed., 1973. *Geothermal Energy*. 185 pp. The Unesco Press, Paris.

Banwell, C. J., 1973. Geophysical methods in geothermal exploration. In *Geothermal Energy*. 41–48. H. C. H. Armstead, ed. The Unesco Press, Paris.

Barth, T. F. W., 1950. *Volcanic Geology: Hot Springs and Geysers of Iceland, Publ 587*. 174 pp. Carnegie Institution of Washington, Washington, D. C.

Benseman, R. F., 1965. The components of a geyser. *New Zealand J. Sci.* **8,** 24–44.

Birch, F. and G. C. Kennedy, 1973. Notes on geyser temperatures in Iceland and Yellowstone National Park. In *Flow and Fracture of Rocks, Geophys. Monogr. Ser. 16*. 329–336. H. C. Heard, I. Y. Borg, N. L. Carter, and C. B. Raleigh, eds. AGU, Washington, D. C.

Birsic, R. J., 1976. *More About Geothermal Steam*. 172 pp. R. J. Birsic, Fullerton, CA.

Bloss, F. D. and T. F. W. Barth, 1949. Observations on some Yellowstone geysers. *Geol. Soc. Amer. Bull.* **60,** 861–886.

Brannock, W. W., P. F. Fix, V. P. Gianalla, and D. E. White, 1948. Preliminary geochemical results at Steamboat Springs, Nevada. *AGU Trans. 29,* **2,** 211–226.

Brock, T. D., 1978. *Thermophilic Microorganisms and Life at High Temperatures*. 465 pp. Springer-Verlag, New York.

Bryan, T. S., 1979. *The Geysers of Yellowstone*. 225 pp. Colorado Associated University Press, Boulder, CO.

Buchtala, J., 1928. Der Geysir von Herlány: Einheitliche Erklärungen der verschiedenen Geysirphänomene en einem aërohydrodynamischen Apparate. *Ztschr. f. prakt. Geol.* **36,** 149–154.

Budd, C. F., Jr., 1975. Steam production at The Geysers geothermal field. In *Geothermal Energy.* 129–144. P. Kruger and C. Otte, eds. Stanford University Press, Stanford.

Bullard, E., 1973. Basic theories. In *Geothermal Energy.* 19–29. H. C. H. Armstead, ed. The Unesco Press, Paris.

Bunsen, R., 1848. On the intimate connection existing between the pseudo-volcanic phenomena of Iceland. In *Memoir VIII, Chemical Reports and Memoirs.* (English translation.) 323–370. T. Graham, ed. Cavendish Memoirs, London.

Carr, W. J. and W. D. Quinlivan, 1966. *Geologic Map of the Timber Mountain Quadrangle, Nye County, Nevada.* U. S. Geol. Surv. Quad. Map GQ-503.

Dana, J. D., 1894. *Manual of Geology.* 4th ed. 1087 pp. American Book Co., New York.

Dawson, G. B., 1964. The nature and assessment of heat flow from hydrothermal areas. New Zealand J. of Geol. and Geophys. **7** (1), 155–171.

Day, A. L. and E. T. Allen, 1925. *The Volcanic Activity and Hot Springs of Lassen Peak, Publ 360.* 190 pp. Carnegie Institution of Washington, Washington, D. C.

Dench, N. D., 1973. Well measurements. In *Geothermal Energy.* 85–96. H. C. H. Armstead, ed. The Unesco Press, Paris.

Eaton, G. P., R. L. Christiansen, H. M. Iyer, A. M. Pitt, D. R. Mabey, H. R. Blank, Jr., I. Zietz, and M. E. Gettings, 1975. Magma beneath Yellowstone National Park. *Science* **188,** 787–796.

Einarsson, S. S., 1973. Geothermal district heating. In *Geothermal Energy.* 123–134. H. C. H. Armstead, ed. The Unesco Press, Paris.

Einarsson, T., 1967. *The Great Geysir and Hot Spring Area of Haukadalur, Iceland.* 24 pp. The Geysir Committee, Reykjavik, Iceland.

Fenner, C. N., 1936. Bore-hole investigations in Yellowstone Park. *J. Geol.* **44,** 225–315.

Fischer, W. A., 1960. Yellowstone's living geology. In *Yellowstone Nature Notes 33.* 62 pp. Yellowstone Library and Museum Association, Yellowstone National Park, WY.

Fournier, R. O. and J. J. Rowe, 1966. Estimation of underground temperatures from the silica content of water from hot springs and wet-steam wells. *Amer. J. Sci.* **264,** 685–697.

Fournier, R. O. and A. H. Truesdell, 1973. An empirical Na-K-Ca geothermometer for natural waters. *Geochim. et Cosmochim. Acta* **37,** 1255–1275.

Gippenreiter, V., 1969. *On the way to Kamchatka's Volcanoes.* (In English and Russian.) 184 pp. Planeta, Moscow, USSR.

Golovina, I. F. and N. N. Malov, 1960. On the geyser theory. *AGU Izv. Geophys. Ser.* Sept. 1960, 922–929. (Engl. trans.)

Grew, J. C., 1925. Waimangu and the hot-spring country of New Zealand. *Natl. Geog. Mag.* **48** (2), 110–130.

Grindley, G. W., 1965. The geology, structure, and exploitation of the Wairakei geothermal field, Taupo, New Zealand. *New Zealand Geol. Surv. Bull.* **N.S. 75,** 131 pp.

Hague, A., 1892. *Folios of the Geologic Atlas of the United States # 30 Yellowstone National Park, Wyo.* 6 pp., 3 illus., 8 maps. U. S. Geol. Surv., Washington, D. C.

Hague, A., 1904a. *Atlas to Accompany Monograph XXXII on the Geology of the Yellowstone National Park.* U. S. Geol. Surv., Washington, D.C.

Hague, A., 1904b. The Yellowstone National Park. *Scribner's* **35**, 513–527.

Hammond, A. L., 1972. Geothermal energy: An emerging major resource. *Science* **177**, 978–980.

Healy, J., 1953. Preliminary account of hydrothermal conditions at Wairakei, New Zealand. *Proc. 8th Pacific Sci. Congr., Philippines* **2**, 214–227.

Healy, J., 1964. Volcanic mechanisms in the Taupo volcanic zone. *New Zealand J. of Geol. and Geophys.* **7**, 151–157.

Hochstetter, F. C. von, 1867. *New Zealand, Its Physical Geography, Geology, and Natural History.* (English translation.) 515 pp. Cotta, Stuttgart, West Germany.

Honda, K. and T. Terada, 1906. On the geyser in Atami. *Phys. Rev.* **22**, 300–311.

Hutchinson, R. W., 1979. Private communicatons.

Iwasaki, I., 1962. Geochemical investigations of geysers in Japan. *Bull. Tokyo Inst. Technol.* **46**, 60 pp.

Keefer, W. R., 1972. The geologic story of Yellowstone National Park. *U. S. Geol. Surv. Bull, 1347,* 92 pp.

Kimura, K., 1949. Geochemical studies on the radioactive springs in Japan. *Proc. 7th Pacific Sci. Congr., Auckland, N. Z.* **2**, 485–489.

Krug von Nidda, C., 1837. On the mineral springs of Iceland. *Edinburgh Philos. Mag. and J. Sci.* **22**, 90–110; 220–226.

Kruger, P. and C. Otte, eds., 1973. *Geothermal Energy.* 360 pp. Stanford University Press, Stanford.

Langford, N. P., 1871. The wonders of Yellowstone. *Scribner's* **1; 2**, 1–17; 113–128.

Lee, W. H. K., ed., 1965. *Terrestrial Heat Flow, Geophys. Monogr. Ser. 8.* 276 pp. AGU, Washington, D. C.

Lindal, B., 1973. Industrial and other applications of geothermal energy. In *Geothermal Energy.* 135–148. H. C. H. Armstead, ed. The Unesco Press, Paris.

Lloyd, E. F., 1965. Whakarewarewa hot springs. In *New Zealand Volcanology—Central Volcanic Region.* 32–37. B. N. Thompson, L. O. Kermode, and A. Ewart, eds. New Zealand Dept. of Sci. and Indust. Res. Inform. Ser. no. 50, Wellington, N. Z.

Lloyd, E. F., 1975. Geology of Whakarewarewa hot springs. *New Zealand Dept. of Sci. and Indust. Res. Inform. Ser. no. 111.* 24 pp.

Lloyd, E. F. and R. F. Keam, 1974. Trinity Terrace hydrothermal eruption, Waimangu, New Zealand. *New Zealand J. of Sci.* **17**, 511–528.

Longman, I. M., 1959. Formulas for computing the tidal acceleration due to the moon and the sun. *J. Geophys. Res.* **64**, 2351–2355.

Lystrup, H. T., 1933. Winds—Their effect on geyser activity. In *Yellowstone Nature Notes 10.* 26 pp. Yellowstone Library and Museum Association, Yellowstone National Park., WY.

Mackenzie, G. S., 1842. *Travels in Iceland*. Revised edition. 88 pp. Chambers, Edinburgh, Scotland.

Mal'tsev, L. M. and A. L. Bubol'ts, 1961. Well-geyser on the Boja-Dag. *Akad. Nauk Turkmen SSR Izv., Ser. Fiz-Tekh., Khim-i Geol. Nauk* **1**, 68−73.

Marler, G. D., 1951. Exchange of function as a cause of geyser irregularity. *Amer. J. Sci.* **249**, 329−342.

Marler, G. D., 1954. Does the cold of winter affect the thermal intensity of the hot springs of Yellowstone Park? *Amer. J. Sci.* **252**, 38−54.

Marler, G. D., 1956. How old is Old Faithful Geyser? *Amer. J. Sci.* **254**, 615−622.

Marler, G. D., 1964a. *Studies of Geysers and Hot Springs Along the Firehole River, Yellowstone National Park, Wyoming*. 49 pp. Yellowstone Library and Museum Association, Yellowstone National Park, WY.

Marler, G. D., 1964b. Effects of the Hebgen Lake earthquake of August 17, 1959, on the hot springs of the Firehole geyser basins, Yellowstone National Park. *U. S. Geol. Surv. Prof. Pap. 435-Q*, 185−197.

Marler, G. D., 1973. *Inventory of Thermal Features of the Firehole River Geyser Basins and Other Selected Areas of Yellowstone National Park*. 648 pp. Natl. Tech. Inf. Service Rept. PB-221289, Washington, D. C.

Marler. G. D., 1974. *The Story of Old Faithful geyser*. Revised edition. 49 pp. Yellowstone Library and Museum Association, Yellowstone National Park, WY.

Marler, G. D. and D. E. White, 1975. Seismic geyser and its bearing on the origin and evolution of geysers and hot springs of Yellowstone National Park. *Geol. Soc. Amer. Bull.* **86**, 749−759.

Martinez, S. J., 1976. Geyser Activity at the Crystal Geyser Area, Green River, Utah. 11 pp. Private communication.

Miller, J. P., 1952. A portion of the system calcium carbonate−carbon dioxide-water. *Amer. J. Sci.* **250**, 161−203.

Milora, S. L. and J. W. Tester, 1976. *Geothermal Energy as a Source of Electric Power*. 186 pp. MIT Press, Cambridge, Mass.

Morey, G. W., R. O. Fournier, and J. J. Rowe, 1962. The solubility of quartz in water in the temperature interval from 25 to 300°C. *Geochim. et Cosmochim. Acta* **26**, 1029−1043.

Morrison, R., P. Saint, and D. Weaver, 1977. Surface disposal of geothermal brines. In *Geothermal: State of the Art, Trans.* 229−230. Geothermal Resources Council, Davis, CA.

Munby, A. E., 1902. A model geyser. *Nature* **65**, 247.

Murbarger, N., 1956. Geysers of Whirlwind Valley. *Desert Mag.* **19** (1), 17−20.

Nekhoroshev, A. S., 1959. On the question of the theory of the functioning of geysers. *Acad. Sci. USSR* **127** (5), 1096−1098. (English translation.)

Nicholls, H. R. and J. S. Rinehart, 1967. Geophysical study of geyser action in Yellowstone National Park. *J. Geophys. Res.* **72**, 4651−4663.

Nogoshi, M. and Y. Motoya, 1962. Tremors at the Onikobe geyser (with special relation to volcanic tremors). *Geophys. Bull. Hokkaido Univ.* **9**, 67−76.

Noguchi, K., 1956. Geochemical investigation of geysers in Japan. *Proc. 8th Pacific Sci. Congr., Philippines* **2**, 228−240.

Nolan, T. B. and G. H. Anderson, 1934. The geyser area near Beowawe, Eureka County, Nevada. *Amer. J. Sci.* **27,** 215–229.

Nomura, Y., 1954. Studies on geysers at Onikobe, Japan. *Tech. Rep. Tohoku Univ.* **19,** 45–62.

Otte, C., 1979. Developing our geothermal energy. *Engr. and Sci. March-April 1979,* 16–22.

Peale, A. C., 1883. The thermal springs of Yellowstone National Park. In *U. S. Geol. and Geogr. Surv. Terr. 12th Ann. Rep. 2.* 63–426. U.S. Government Printing Office. Washington, D. C.

Peterson, R. E. and N. El-Ramly, 1976. The worldwide electric and nonelectric geothermal industry. In *Geothermal World Directory, 1975/76.* 167–176. K. F. Meadows, ed. K. F. Meadows, Glendora, CA.

Rinehart, J. S., 1965. Earth tremors generated by Old Faithful geyser. *Science* **150,** 494–496.

Rinehart, J. S., 1968a. Seismic signatures of some Icelandic geysers. *J. Geophys. Res.* **73,** 4609–4614.

Rinehart, J. S., 1968b. Geyser activity near Beowawe, Eureka County, Nevada. *J. Geophys. Res.* **73,** 7703–7706.

Rinehart, J. S., 1969a. Thermal and seismic indications of Old Faithful geyser's inner workings. *J. Geophys. Res.* **74,** 566–573.

Rinehart, J. S., 1976b. *A Guide to Geyser Gazing.* 64 pp. HyperDynamicS, Santa Fe, NM.

Rinehart, J. S., 1970a. Heat flow from natural geysers. *Tectonophys.* **10,** 11–17.

Rinehart, J. S., 1970b. Geysering action in a drilled well, Crump, Lake County, Oregon. *J. Geophys. Res.* **75,** 6714–6716.

Rinehart, J. S., 1972a. Fluctuations in geyser activity caused by variations in earth tidal forces, barometric pressure, and tectonic stresses. *J. Geophys. Res.* **77,** 342–350.

Rinehart, J. S., 1972b. 18.6-year tide regulates geyser activity. *Science* **177,** 346–347.

Rinehart, J. S., 1973. Change in Old Faithful's activity before the March 1973 earthquakes (abstract). *EOS Trans., AGU* **54,** 1142.

Rinehart, J. S., 1974. Geysers. *EOS Trans., AGU* **55,** 1052–1062.

Rinehart, J. S., 1975. Geothermics. *Appl. Mech. Rev.* **28,** 1081–1084.

Rinehart, J. S., 1976a. Geysers and geothermal power production. *Die Naturwiss.* **63,** 218–223.

Rinehart, J. S., 1976b. *A Guide to Geyser Gazing.* 64 pp. HyperDynamicS, Santa Fe, NM.

Rowe, J. J., R. O. Fournier, and G. W. Morey, 1965. Use of sodium iodide to trace underground water circulation in the hot springs and geysers of the Daisy group, Yellowstone National Park. *U. S. Geol. Surv. Prof. Pap. 525-B,* 184–186.

Rowe. J. J., R. O. Fournier, and G. W. Morey, 1973. Chemical analysis of thermal waters in Yellowstone National Park, Wyoming, 1960–65. *U. S. Geol. Surv. Bull. 1303,* 31 pp.

Sigvaldason, G. E. and D. E. White, 1961. Hydrothermal alteration of rocks in two drill holes at Steamboat Springs, Washoe County, Nevada. *U. S. Geol. Surv. Prof. Pap. 424-D,* 116–122.

Stacey, F. D., 1969. *Physics of the Earth*. 324 pp. John Wiley, New York.

Stearns, N. D., 1933. A remarkable intermittent spring. *Mid-Pacific Mag.* **45** (3), 216–218.

Steinberg, G. S., A. G. Merzhanov, and A. S. Steinberg, 1978. Hydrosounding as a method of study of the critical parameters of the geysers. *J. Volcan. Geotherm. Res.* **3**, 99–119.

Tester, J. W., G. E. Torris, R. G. Cummings, and R. L. Bivins, 1979. *Electricity from Hot Dry Rock Geothermal Energy: Technical and Economic Issues. LA-7603-MS; UC-66f.* 24 pp. University of California, Los Alamos Scientific Laboratory, Los Alamos, NM.

Thompson, G. A. and D. B. Burke, 1974. Regional geophysics of the Basin and Range Province. In *Annual Review of Earth and Planetary Sciences*. 213–238. Annual Reviews, Inc., Palo Alto, CA.

Thompson, J. M., T. S. Presser, R. B. Barnes, and D. B. Bird, 1975. Chemical analysis of the waters of Yellowstone National Park. *U. S. Geol. Surv. Open-File Rep. 75-25,* 59 pp.

Thompson, J. M. and S. Yadav, 1979. Chemical analysis of waters from geysers, hot springs and pools in Yellowstone National Park, Wyoming, from 1974 to 1978. *U. S. Geol. Surv. Open-File Rep. 79-704,* unpaged.

Thorkelsson, Th., 1910. The hot springs of Iceland. *Kgl. Danske Vidensk. Selsk Skr, Ser. 7, Sec. 8, No. 4,* 181–264.

Thorkelsson, Th., 1928a. On the geyser theory. *Philos. Mag.* **5,** 441–444.

Thorkelsson, Th., 1928b. *On Thermal Activity in Reykjanes, Iceland.* 42 pp. Societas Scientiarum Islandica, No. 3, Reykjavik, Iceland.

Thorkelsson, Th., 1940. *On Thermal Activity in Iceland and Geyser Action.* 139 pp. Isafoldarpretsmidja, Reykjavik, Iceland.

Truesdell, A. H. and R. O. Fournier, 1975. Conditions in the deepest parts of the hot spring system of Yellowstone National Park, Wyoming. *U. S. Geol. Surv. Open-File Rep.*

Truesdell, A. H. and R. O. Fournier, 1976. Calculation of deep temperature in geothermal systems from the chemistry of boiling spring waters of mixed origin. In *Proc. UN Symposium on Development and Use of Geothermal Resources.* 837–844. U.S. Government Printing Office, Washington, D. C.

Truesdell, A. H. and R. O. Fournier, 1977. Procedure for estimating the temperature of a hot-water component in a mixed water by using a plot of dissolved silica versus enthalpy. *J. Res., U. S. Geol. Surv.* **5,** 49–52.

Tyndall, J., 1851–1854. On some of the eruptive phenomena of Iceland. *Proc. Roy. Inst. Great Britain* **1,** 329–335.

Tyndall, J., 1895. *Heat: A Mode of Motion.* 6th ed. 166–173. D. Appleton, New York.

Ustinova, T. I., 1955. Kamchatka geysers. *Geographgiz, Moscow USSR.* 120 pp.

Verhoogen, J., F. J. Turner, L. E. Weiss, C. Wahrhaftig, and W. S. Fyfe, 1970. *The Earth, An Introduction to Physical Geology.* 748 pp. Holt, Rinehart, and Winston, New York.

Vymorokov, B. M., 1960. How geysers work. *Priroda* **11,** 97–99.

Wahl, E. F., 1977. *Geothermal Energy Utilization*. 302 pp. John Wiley, New York.

Waring, G. A., 1965. Thermal Springs of the United States and other countries of the world—A summary. *U. S. Geol. Surv. Prof. Pap. 492*, 383 pp. (Revised by R. R. Blankenship and R. Bentall).

White, D. E., 1965. Geothermal energy. *U. S. Geol. Surv. Circ. 519*, 17 pp.

White, D. E., 1967. Some principles of geyser activity mainly from Steamboat Springs, Nevada. *Amer. J. Sci.* **265**, 641−684.

White, D. E., 1968. Hydrology, activity and heat flow of the Steamboat Springs thermal system, Washoe County, Nevada. *U. S. Geol. Surv. Prof. Pap. 458-C,* 109 pp.

White, D. E. and W. W. Brannock, 1950. The sources of heat and water supply of thermal springs, with particular reference to Steamboat Springs, Nevada. *AGU, Trans.* **31,** 569−572.

White, D. E. and D. L. Williams, eds., 1975. Assessment of geothermal resources of the United States. *U. S. Geol. Surv. Circ. 726,* 155 pp.

White, D. E., W. W. Brannock, and K. J. Murata, 1956. Silica in hot-spring waters. *Geochim. et Cosmochim. Acta* **10,** 27−59.

White, D. E., R. O. Fournier, L. J. P. Muffler, and A. H. Truesdell, 1975. Physical results of research drilling in thermal areas of Yellowstone National Park, Wyoming. *U. S. Geol. Surv. Prof. Pap. 892,* 70 pp.

Index